AF696454

HAND-TO-HAND COMBAT

THE V-FIVE PHYSICAL EDUCATION SERIES

THE UNITED STATES NAVAL INSTITUTE is a nationwide organization of military and civilian members. The Institute was founded in 1873, not for profit, but for the advancement of professional, literary, and scientific knowledge in the Navy and among military and civilian contemporaries.

Maintenance of high standards of physical fitness in the youth of our nation is considered a prerequisite to national preparedness. With this in mind, the Naval Institute makes a continuing effort to insure revisions, through the V-Five Association of America, that are compatible with civilian educational interests.

The V-Five Physical Education Series

HAND-TO-HAND COMBAT

Annapolis, Maryland

UNITED STATES NAVAL INSTITUTE

Hand-to-Hand Combat

Copyright © 1943, 1970 by the United States Naval Institute

ISBN 1-58160-423-8
Printed in the United States of America

Reprint edition published by Paladin Press, a division of
Paladin Enterprises
Gunbarrel Tech Center
7077 Winchester Circle
Boulder, CO 80301 USA
+1.303-443-7250

Direct inquiries and/or orders to the above address.

PALADIN, PALADIN PRESS, and the "horse head" design
are trademarks belonging to Paladin Enterprises and
registered in the United States Patent and Trademark Office.

All rights reserved by the United States Naval Institute.
Except for use in a review, no portion of this book may
be reproduced in any form without express written permission
by the United States Naval Institute.

Neither the author, publisher, nor copyright holder assumes
any responsibility for the use or misuse of information
contained in this book.

Paladin Enterprises, Inc., is in no way affiliated with or sponsored
by the United States Naval Institute.

Visit the Paladin Press Web site at www.paladin-press.com

Visit the United States Naval Institute Web site at www.usni.org

Preface

Sports are used as a training device in the physical training program for Naval Aviation just as mathematics and physics are taught in the academic courses and ordnance and gunnery are employed in the military education of cadets. Physical training was made an integral part of the training plan and is continued progressively throughout the entire training of aviation cadets. Successful coaches were commissioned so that the Navy might have the best instruction available.

Each sport has definite objectives of its own, and, in addition, contributes to the over-all aims and purposes of Naval Aviation training. In such a program it is natural that at times experience may show better means of achieving the desired objectives. Initially, syllabi were prepared by the newly commissioned athletic officers to serve as guides in conducting the various sports and activities. While adhering closely to the original plans, the experience gained in teaching thousands of cadets in varying circumstances, has been a valuable supplement to the physical training program. The basic features of the original program, plus the results of a year of training, are now published in this series, The Naval Aviation Physical Training Manuals. These manuals have been prepared by and for the officers in charge of the physical training of Naval Aviation personnel.

In any consideration of the use of these manuals, it is important to understand that in the Naval Aviation program, sports are not used for their own sake or for recreational purposes. The competitive sports embodied in this program were selected for what they contribute to the development of desirable characteristics in the aviation cadet. It is for this reason that the books are unlike other sports publications. The manuals consider sports in the military training sense, in their conditioning values, both mental and physical.

The daily story of the war emphasizes again and again the fact that we are facing an enemy which is careless of life because they are so steeped in a fanatical nationalism. The common rules of war mean nothing to a desperate enemy. It is our duty to train the cadets to be superior to that enemy, mentally and physically. Rigorous, tough, competitive sports offer an excellent medium to fulfill this mission. Records have proven that mental improvement of the cadets go hand in hand with better physical condition.

It is the duty of each athletic officer, through observation and through the study of these manuals, to familiarize himself with all the sports and activities in this training program. At any time he may be assigned to instruct in any sport or he may be designated to administer a complete physical training program.

This manual has been prepared by the officers in charge of the instruction of Hand-to-Hand Combat in Naval Aviation.

A. W. RADFORD
Captain, U.S.N.
Director of Training
Bureau of Aeronautics, U. S. Navy

April, 1943

Introduction

HAND-TO-HAND COMBAT is an essential part of modern warfare. Various methods, or systems, of close-quarter fighting have been urged as the most practical way to kill or overcome an enemy in personal encounter. Commando tactics, the jiu jitsu, boxing, wrestling, and other systems, all have their proponents. Hand-to-hand combat makes use of the deadliest maneuvers of each of these forms of fighting, and adds forms of attack not contained in any of them.

In other words, hand-to-hand combat does not replace any known system of close fighting. Instead, it makes use of all known forms of personal combat, and any other means that will accomplish a quick kill. There is nothing glamorous or adventurous about it, as one often finds in instruction books on Commando training. It lacks the mystery surrounding jiu jitsu tactics. It simply is a cold, efficient method of overcoming your enemy in a manner most suitable to the performance of your mission or the saving of your life. Whether it is to your advantage to silence him quickly and take him prisoner, or to kill him outright, the training teaches the simplest and most efficient method of accomplishing your task—and without any regard to the comfort or pleasure of the enemy.

It is necessary that Naval personnel be trained in hand-to-hand combat. Already, in World War II Navy men have been in hand-to-hand combat in many places. The widespread areas over which this war is being fought indicate the continued probability of men being forced down behind enemy lines, or being placed in position where their most effective weapons are those provided them by nature. Close contact with sentries, small patrols, or isolated enemy detachments is to be expected when making an effort to pass through enemy lines. Efficient hand-to-hand combat tactics usually decide the answer to whether or not you get through.

Hand-to-hand combat has but one simple objective. That objective is to *Win*. There is no prize for finishing second in a hand-to-hand fight with the enemy. You kill or capture your opponent as quickly as possible. There are no rules; there is nothing "sporting" about it. No holds are barred, no fouls are considered, and there is never an umpire or a referee. You do to the enemy exactly what he would like to do to you—only, you do it first!

On the pages of this manual will be found illustrations of the best possible tactics and techniques. Actual experience has demonstrated what methods have served their purposes best. Practicability was the thought uppermost in the minds of the authors in preparing this manual; not the spectacular, the mysterious, or the complex.

The instructor or student may wonder why certain maneuvers were not included in this manual. They were excluded either because of some defect in their application, or because of great complexity in their movement. It is the purpose of this manual to save lives instead of to experiment with them, and at the same time to give the fighting man a knowledge of the deadly weapons he possesses within his own body.

Table of Contents

HAND-TO-HAND COMBAT

CHAPTER I

History

HAND-TO-HAND COMBAT is not new. Some phases of its development are due to changes in weapons, yet its basic principles have remained the same for many years. In centuries past similar tactics were known more familiarly as guerrilla warfare, a type of operation in which individual contact was the rule and where success or failure of an attack hinged on the results of hand-to-hand encounters. The battles of the ancient Persians, Greeks, and Romans are good examples of such combat. Many difficult missions, even in modern war, frequently are carried out by individuals and by small groups. The British Commandoes, the American Rangers, and the infiltration tactics of the Japanese in Malaya and Burma are cases in point.

Hand-to-hand combat does not evolve about any single style of defense and counter attack. In the past many methods and styles were developed. These include boxing, wrestling, savate, football, fencing, and jiu jitsu. All of the aforementioned include death dealing blows, thrusts or holds which, if applied properly and quickly, have devastating results. Obviously none can wholly be adapted to warfare tactics because they will not stand up under extreme conditions. Hand-to-hand combat adopts the most effective maneuvers from each.

Let us look briefly at these physical combat styles. Boxing, for example, is the art of fighting with the clenched fists covered with padded gloves and permits no striking below the belt, back hand blows, or the "rabbit punch." Thus it has its set of restrictions and is designed to defeat an adversary in a certain manner. For hand-to-hand combat, it might well be instructive for the pupil to study the techniques employed by the ancient Greeks and Romans when gloves of metal, studded with spikes at the knuckles, were used. This would compare with "brass knuckles" used by brawlers today. In hand-to-hand combat this would be very effective, because there is no such thing as a "foul." A blow below the belt, on the nape of the neck, or into the kidneys is part of the fight.

Wrestling is an ancient sport and was popular with the early Romans and Greeks. Modern wrestling has developed along three different lines—Graeco-Roman, the combined style of the Ancients, now confined to Europe; catch-as-catch-can, or American style, practiced in all universities and high schools; jiu jitsu, said to have originated in Japan, but actually taken from the Lama monks of China. Wrestling generally consists either of pinning an opponent's shoulder to the earth or throwing him from his feet. A variety of trips and holds employing the use of leverage and balance are used along with speed to accomplish an opponent's defeat. Most of us have observed how many holds and maneuvers have been barred from wrestling. Why? Because they were dangerous to the participants and often caused serious injuries. The "twisting hammerlock," the "full nelson" and the "twisting elbow lock" are of this type. Because of the danger and chances of injury, these combinations are exceedingly valuable to hand-to-hand combat.

Savate, a French form of combat, employs the feet, which are normally covered with light leather shoes. Thus, the participant, using the same technique of a boxer, strikes out at his opponent with either hand or foot. Many a knockout has occurred when a foot was planted expertly under the jaw of an adversary. Kicks to the groin, illegal in savate, are valuable in hand-to-hand, as are short upward or downward jabs of the knees.

The game of football can be traced back to the ancient Spartans. Today football embraces three styles of play—soccer, rugby (a modification of soccer), and American football (a combination of both soccer and rugby). We prefer to think of football as played in America. Rules and types of clothing have tended to make it a comparatively safe sport, but clipping and other illegal maneuvers in football are real assets to hand-to-hand.

For centuries fencing has been both a means of deadly combat and a sport. It has been practiced by all types of people, not only to develop quickness and alertness afoot and by hand, but as a means of protection. It has employed all types of swords and daggers. It was an art which brought sudden death to the lesser skilled adversary. A student of hand-to-hand combat should learn some of the techniques. Fencing alone will not make his life secure, but if he combines it with other tactics a knowledge of fencing will prove effective in hand-to-hand fighting.

Finally, we have jiu jitsu, Japanese wrestling, supposedly mysterious and endowed with a so-called power of overcoming any type of opponent. It is neither mysterious nor is it endowed with any superhuman possibilities. The American need become only an average catch-as-catch-can wrestler to overcome ninety-nine out of every hundred jiu jitsu experts. This, of course, applies to a "free for all" or rough and tumble fight. Jiu jitsu mystery comes from the use of Japanese terms to describe certain parts of the body and various maneuvers. In reality these terms can be explained in simple English. The black belt (jiu jitsu) men spend years learning maneuvers and are impressed with the overpowering results they acquire.

The Lama monks of China developed the jiu jitsu form of combat to protect themselves, though unarmed, from the armed robbers on the desolate roads of old China. The Japanese, being small and short in stature, liked it and developed it further. Today every school boy in Japan learns jiu jitsu and practices it throughout his life.

There are many maneuvers which have a practical value in hand-to-hand combat. Jiu jitsu alone would never suffice, even though a great emphasis is placed on leverage, foot action, and body movement to overcome the superior weight of an adversary. However, many of the standing throws and trips are excellent, and can be combined with our other sciences.

As you study this manual you will see how hand-to-hand combat employs the maneuvers and tactics of every type of physical combat, using them to their greatest advantage. A wrestling maneuver may be combined with a boxing blow; jiu jitsu may assist in the delivery of another blow; wrestling with savate; football with boxing; fencing and boxing, or any combination of two or more types.

Hand-to-hand combat not only uses all physical combat techniques, but develops new ones. It teaches methods of disarming an opponent who may be armed with any type of firearm, knife, sword, club, or dangerous weapon, such as a chair, rock, bottle, or the like. It teaches the fighting man how best to protect himself and make the best use of any of these weapons.

Proper methods of searching prisoners are taught, as are techniques for leading and controlling them under various circumstances. In other words, every possible condition is analyzed and studied, thus giving an effective method for handling all situations in personal combat.

The basic principle to be kept in mind when studying hand-to-hand combat is to progress from the first maneuver to combinations. It is more than applying a single tactic or just enough pressure to gain temporary control as in a wrestling match. To illustrate, let us assume you are in a wrestling match. You attack your opponent and catch him in a leg pick-up. You secure his legs and raise him from the deck. Now go to one knee, drop your opponent on his back or shoulders, and try for a pin hold. Simple, isn't it? And no one gets hurt!

Now the situation has changed. Your opponent is bent on eliminating you from his path, and it's not on any soft mat either. You pick him up the same way, only this time you don't drop to your knee in the process of throwing him to the deck; you bounce him hard on the top of his head or the back of his neck. Follow this up with a downward thrust of your knee with your full weight into his ribs and strike him with your knuckles in the Adam's apple, then kick him either in the face, kidneys, solar plexus or groin, or lock his arm against the joint and break the elbow with a sudden jerk of twist. Thus complete control is quickly obtained and you are free to carry on.

The fighting man who becomes accomplished in the art of hand-to-hand combat has additional and vital means of self preservation at his command. It gives him a superiority over any opponent when he is attacked. In a life-and-death struggle his chances of surviving are tremendously increased. Then, too, it gives him a fighting chance for his life when he knows that to submit to the enemy means certain death. There may be that occasion when it is better to take a chance than to be captured. And if you are captured, a knowledge of hand-to-hand combat may provide you with a means of escape.

CHAPTER 2

Safety Precautions

HAND-TO-HAND COMBAT, unlike physical combative sports, cannot be learned or practiced with the same vigor or forceful action. It must be kept constantly in mind that hand-to-hand is not a sport and that its primary purpose is to injure or kill.

These safety measures should be studied and adhered to constantly.

At the outset, proper safety measures must be taken and the student must be impressed with the danger connected with the application of the maneuvers. At no time should a class be left unsupervised. Many of the tactics employed are so dangerous that movement in practice can be carried out only through easily and lightly struck blows. In all cases where a new maneuver is being taught, it should be worked first at slow motion, then speeded up as the students show improvement.

Students acting as "guinea pigs" never should resist actively while practicing, as most injuries generally result on such occasions. As each new maneuver is presented to a class, the resultant damage it will cause should be clearly pointed out.

After the student is impressed thoroughly with the possibilities of hand-to-hand combat, he should be encouraged to practice the many different maneuvers. However, he should be urged to use caution.

The following safety precautions should always be adhered to during instructional periods:

1. Strict and constant supervision of all classes during instructional period.

2. Constant warning of injuries if tactics are forcefully carried out. Each student while acting as a "guinea pig" or "stooge" should resist the tactic passively. Active resistance results in injuries.

3. The instructor should give a complete illustration of each tactic before allowing the class to use it, pointing out the effects on an opponent under actual combat conditions. At the same time it can be shown how far one can go with a maneuver without causing injury.

4. All students should participate in a three to five minute warm-up whenever they have not taken part in some sports activity the hour previous to the hand-to-hand combat session.

Bending, stretching, tugging, or twisting exercises are desirable. A short wrestling session is excellent if the proper facilities are at hand. If students are reporting to class immediately from mass exercises, no additional exercises are necessary, nor is there any need for special exercises.

If the above precautions are followed, injuries will not result. However, there are occasions when through confusion or lack of balance a student might unwittingly injure another. It therefore is apparent that a doctor should always be available during exercises.

CHAPTER 3

Facilities and Equipment

CLASSES in hand-to-hand combat should be held under conditions which simulate reality as nearly as possible. If the tactics and maneuvers are practiced indoors, floor mats must be provided. When weather permits outdoor practice, it is desirable to conduct training on a sodded ground instead of rough or hard ground. The student, under these circumstances, learns out of necessity to protect himself and is impressed with certain instructional fundamentals.

It is recommended that at least eight square feet of space be provided for each pair of students, thus allowing room for the various throws and trips. It has been noted that injuries are often due to one pair of participants falling on or over another, the result of lack of space rather than the actual maneuver itself. Some tactics do not require the same space as others, and it is up to the instructor to ascertain just what his space requirements are and, if possible, to adjust his class size to the space allowed. Outdoor practice areas relieve this problem, but here, also, the instructor should watch carefully to determine the number of students he can instruct and supervise properly.

Students may report for class either in work dress or gym gear. Although considerable wear and tear is given to regular clothing it helps to simulate actual conditions.

Special equipment is necessary to conduct the class properly. Following is the essential equipment needed for instruction:

1. Rubber, leather or soft plastic knives or daggers. (To simulate knife or dagger attack and defense.) Number should equal one-half total number in the class.

2. Night sticks or wooden clubs. These may be purchased or made from broom handles. Number equal to total in class.

3. Dummy pistols or regular pistols, if obtainable. Many instructors have found it necessary to have rough wood models made. Number equal to one-half total enrolled in class.

4. Dummy rifles, or regular rifles if obtainable. Wood models are often desirable. Number equal to one-half of total enrollment of class.

5. Metal supporters may be obtained for each student although they are not necessary. Extreme caution should be used in practice of blows and kicks to the groin, whether or not supporters or used.

6. Padding for the arms, legs, face, and body may be used, but this is unnecessary if safety precautions are followed.

If possible, the instructor should have available real models of weapons which might be used against the student under combat conditions. The peculiarities of each may be pointed out and observed.

CHAPTER 4

Vulnerable Areas

Structural Weakness of the Human Body Subject to Attack

NO ATTEMPT will be made to teach the student a course in anatomy, yet he should know how vulnerable the human body is to attack. By a vicious attack to certain weaker points of an enemy's body, he can be overcome and subdued easily.

A study of the accompanying charts clearly illustrates all the vital areas and pressure points sensitive to various types of attack whether by hand, fist, finger, knee, foot, or any available weapon. Any attack to these areas will either greatly hamper or completely incapacitate an opponent.

Pressure points (marked ▲) are those areas where veins and arteries are vulnerable to a knife attack. Death will result quickly if the bleeding is not stopped. Finger pressure to these areas will stop bleeding. It is necessary to know these areas, not only for attack, but for self-preservation. Frequently sharp blows with a club or clublike weapon to these susceptible areas will result in hemorrhages under the skin or a partial paralysis of that part of the body attacked.

Vital areas (marked ●) are those areas which, if attacked by hand, knee, foot or various weapons will either kill, seriously injure, or cause excessive pain. Any attack to these areas makes hand-to-hand combat easy.

The implement used or the position of your opponent is not important in an attack. But it is important that you attack a vital area. The foot or knee, for example, to the solar plexus or the groin is just as effective as the fist, and it makes no difference whether an adversary is standing, crawling, on his back, side, or atop you.

A blow delivered to the Adam's apple, the solar plexus, the liver, the groin and the tall bone or coccyx creates a temporary paralysis to the entire body, thus making it easy to incapacitate or annihilate your foe. A blow delivered to the vagus nerve endings behind the carotid arteries in the throat, the thoracic vertebra and the mental foramen nerve on the chin sets up a nerve shock which causes your opponent to lose consciousness.

In warfare the spirit of fair play is forgotten—it's either you or the other fellow. Make use of every available means at your command to subdue him. If you don't, he will subdue you, and it won't be pleasant.

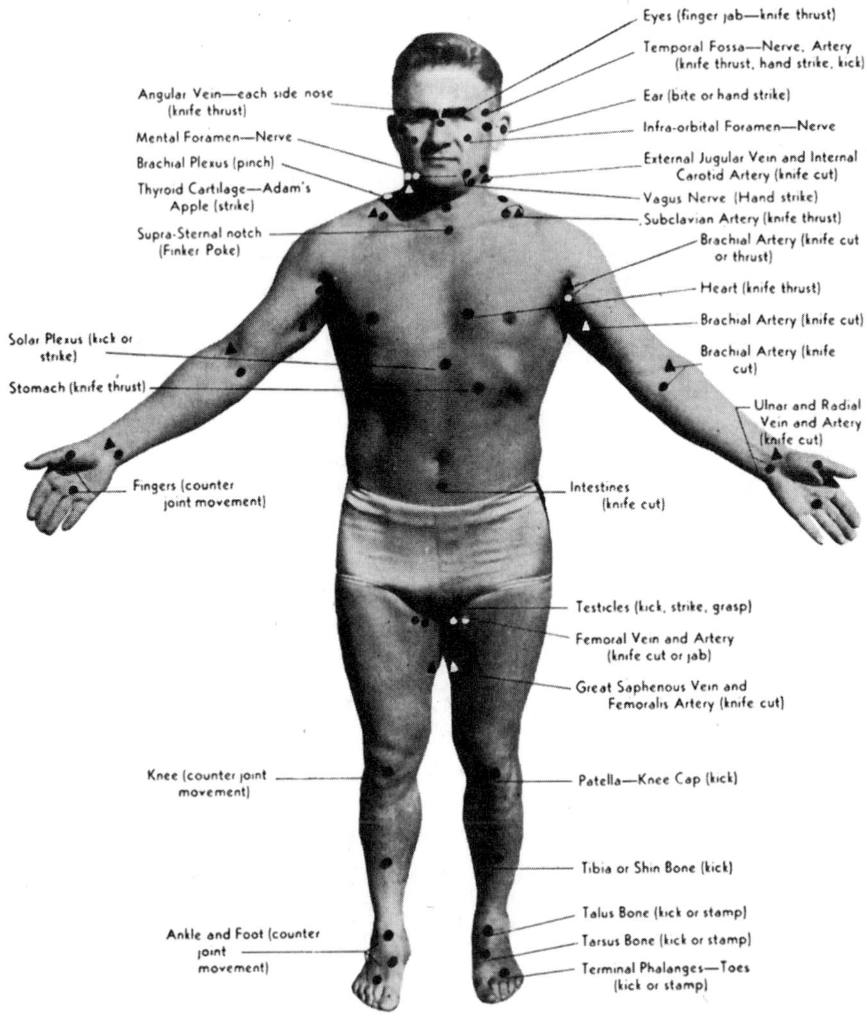
Eyes (finger jab—knife thrust)
Temporal Fossa—Nerve, Artery (knife thrust, hand strike, kick)
Angular Vein—each side nose (knife thrust)
Ear (bite or hand strike)
Mental Foramen—Nerve
Infra-orbital Foramen—Nerve
Brachial Plexus (pinch)
External Jugular Vein and Internal Carotid Artery (knife cut)
Thyroid Cartilage—Adam's Apple (strike)
Vagus Nerve (Hand strike)
Subclavian Artery (knife thrust)
Supra-Sternal notch (Finker Poke)
Brachial Artery (knife cut or thrust)
Heart (knife thrust)
Brachial Artery (knife cut)
Solar Plexus (kick or strike)
Brachial Artery (knife cut)
Stomach (knife thrust)
Ulnar and Radial Vein and Artery (knife cut)
Fingers (counter joint movement)
Intestines (knife cut)
Testicles (kick, strike, grasp)
Femoral Vein and Artery (knife cut or jab)
Great Saphenous Vein and Femoralis Artery (knife cut)
Knee (counter joint movement)
Patella—Knee Cap (kick)
Tibia or Shin Bone (kick)
Talus Bone (kick or stamp)
Ankle and Foot (counter joint movement)
Tarsus Bone (kick or stamp)
Terminal Phalanges—Toes (kick or stamp)

Chart I

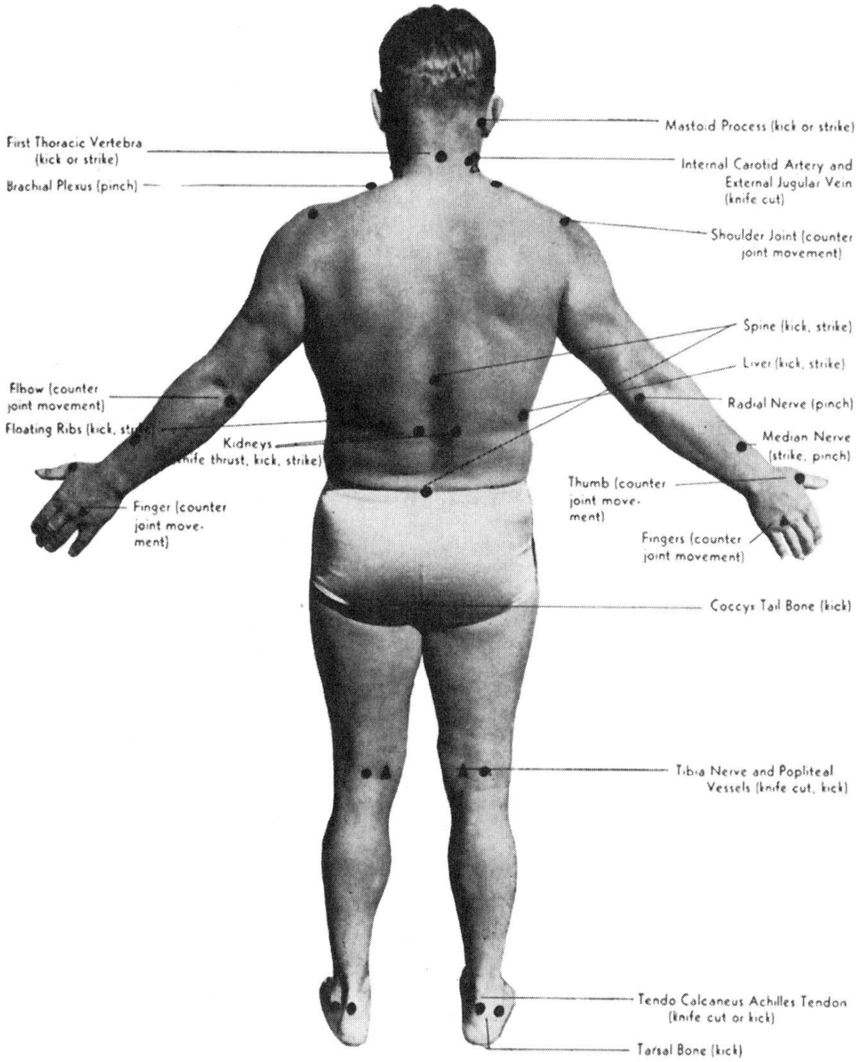
Mastoid Process (kick or strike)
First Thoracic Vertebra (kick or strike)
Internal Carotid Artery and External Jugular Vein (knife cut)
Brachial Plexus (pinch)
Shoulder Joint (counter joint movement)
Spine (kick, strike)
Liver (kick, strike)
Elbow (counter joint movement)
Radial Nerve (pinch)
Floating Ribs (kick, st
Median Nerve (strike, pinch)
Kidneys
nife thrust, kick, strike)
Thumb (counter joint movement)
Finger (counter joint movement)
Fingers (counter joint movement)
Coccyx Tail Bone (kick)
Tibia Nerve and Popliteal Vessels (knife cut, kick)
Tendo Calcaneus Achilles Tendon (knife cut or kick)
Tarsal Bone (kick)

Chart II

CHAPTER 5

Fundamentals

THE INSTRUCTOR and student will find this chapter most helpful to further their study of hand-to-hand combat. Each fundamental is illustrated, its application is described, and additional means of usage are listed. They will be used time and again when actually carrying out the maneuvers and combinations illustrated and explained in later chapters.

It is the purpose in listing these fundamental tactics, not so much to bring new ideas to the mind of the student and instructor, but to refresh their memories in things already known but seldom practiced. Deeply ingrained in American youth is that feeling of sportsmanship and gentlemanly conduct. Today, as we face enemies who recognize no fair play, our techniques of man to man competition must be drastically revised to fit the tactics of war.

As one examines these fundamentals he realizes that the code of sportsmanship is suspended "for the duration." A picture can tell more than a thousand words, and because of the clearness and completeness of the illustrations, it will not be necessary to discuss at length the various techniques offered. Give thought and consideration to the illustrations, learn the fundamentals, and progress through the course will be rapid.

A. Stance

1. HANDS ON HIPS

Be on the alert at all times, hands on hips, legs comfortably apart, one slightly forward, knees very slightly bent. This allows for quick action—bringing hands and arms up to defense, and turning or stepping out of range or to focal point of action.

2. HANDS CROSSED ON CHEST

On the alert, arms folded on chest, legs comfortably apart, one slightly forward. Allows for quick action as illustration 1 above.

In both this stance and in one above, the hands can be brought up, one to each side of the face with the forearms protecting the chest and abdomen. Crouching further with hands and arms in this position places you in position for any type of attack or defense. The position is a combination of the boxer's and wrestler's stance.

3. CONTACT STANCE

Knees are brought quickly together from the positions shown in illustrations 1 and 2 in order to avoid sudden kicks to the groin. Hands and arms can be maneuvered to any desired position.

B. Hand as Weapon

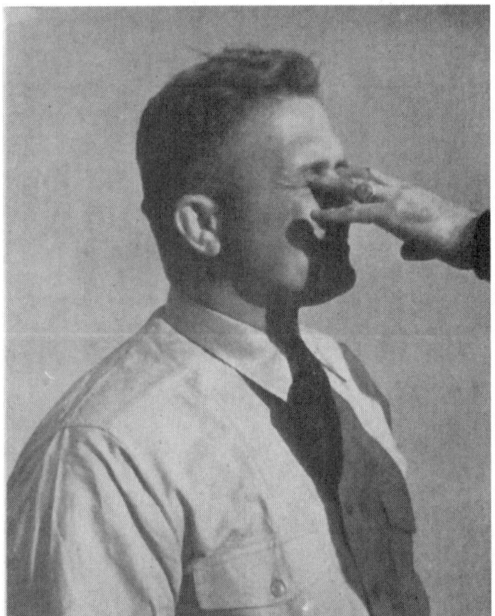

4. FINGER JAB—EYES

The fingers are extended with a sharp thrust to the eyes for a penetrating effect. Fingers are stiff but may be spread slightly, wrist is firm. Dangerous and causes loss of sight. Can be delivered from a variety of positions. In order to gouge out the eyes, insert thumbs or fingers at the outer edge of the eye ball and press inward. With a quick thrust accompanied by a straightening of the finger joint pops the eye out.

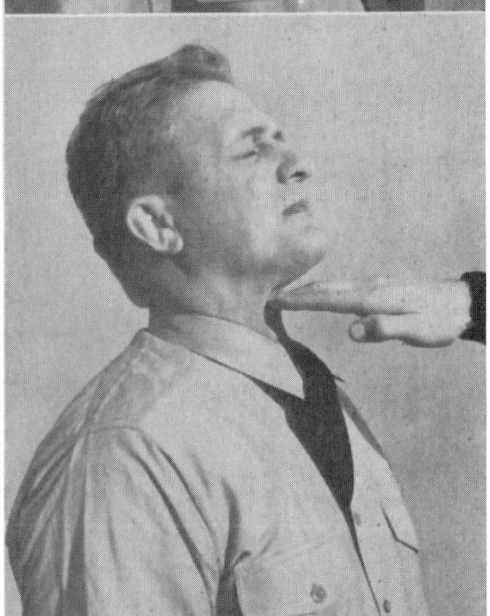

5. FINGER JAB—ADAM'S APPLE

Effective to this part of the throat. Causes temporary loss of speech or paralysis; fracture to voice box may cause death. Very painful.

6. FINGER JAB—SOLAR PLEXUS

Effective to solar plexus. The sharp blow delivered unexpectedly may cause an opponent to drop to knees or bend forward. The jab has a paralyzing effect.

7. KNUCKLE JAB—EYES

Fingers are tucked so that 2nd knuckles are foremost. Sharp blows to the eyeballs causing loss of sight, temporarily. May be delivered from a variety of positions.

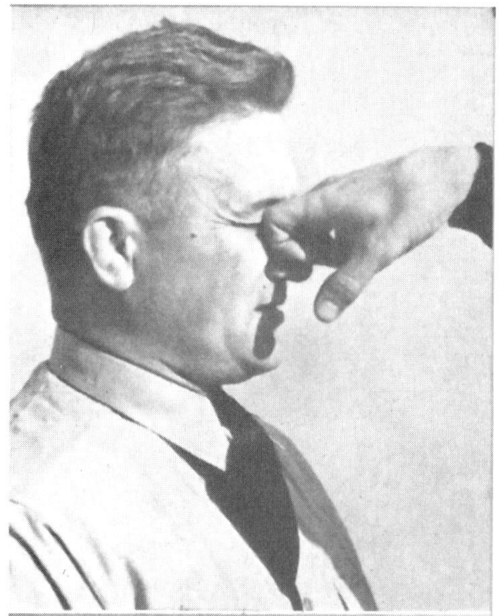

8. KNUCKLE JAB—NOSE

Effective to the nose, especially at the bridge. Causes severe pain and hampers vision.

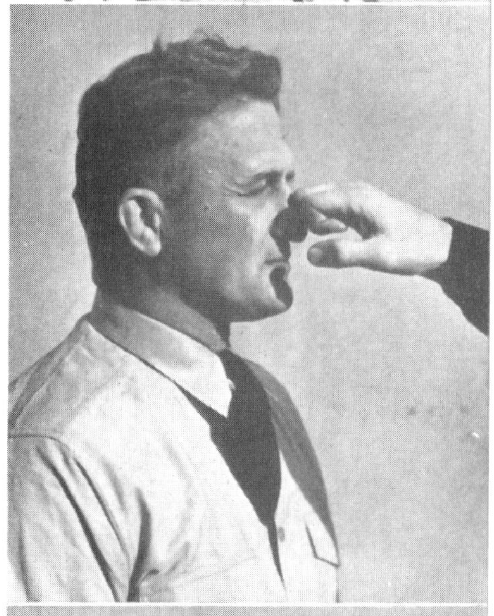

9. KNUCKLE JAB—UPPER LIP

Effective to upper lip. May strike nerves or upper teeth.

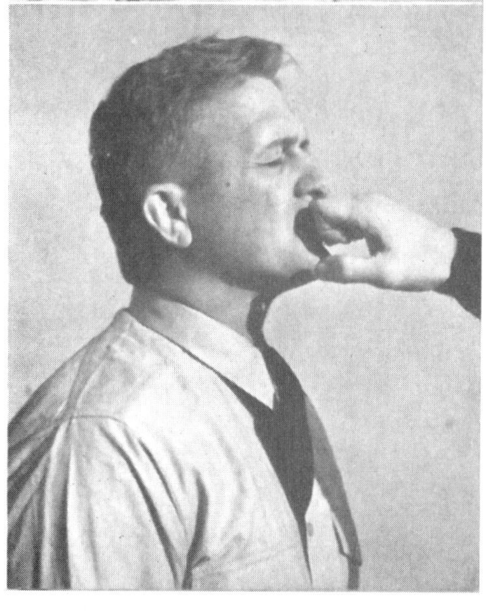

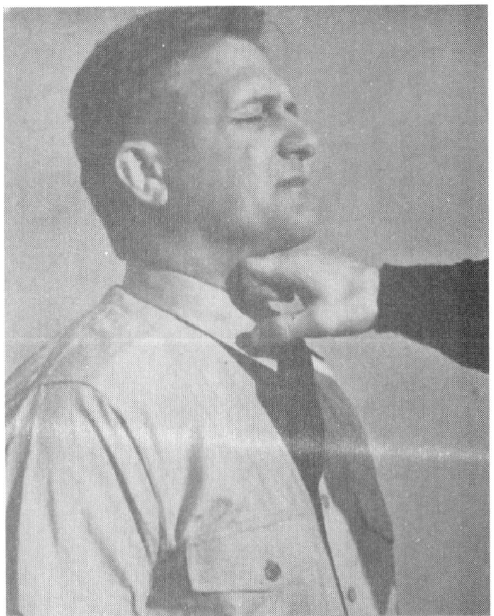

10. KNUCKLE JAB—ADAM'S APPLE

Effective to Adam's apple or larynx. Has the same result as the finger jab, but more severe.

11. KNUCKLE JAB—SOLAR PLEXUS

Effective to solar plexus or stomach. May cause person to bend over quickly or drop to knees.

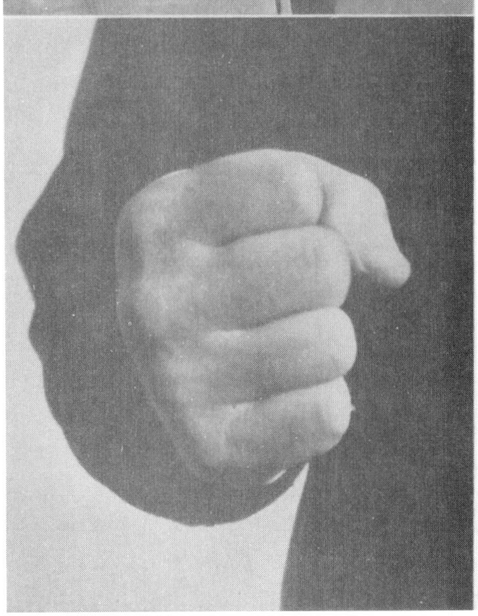

12. FIST—FORMED FIST

Note the proper formation of the fist—the thumb is folded over the outside of the fingers. Should be used only on soft vulnerable body areas. Knuckles and the bones of the hands when unprotected injure easily if striking bony portions of the anatomy.

13. FIST—SOLAR PLEXUS

The fist struck firmly into the solar plexus is an excellent and devastating blow. It will cause an opponent to lose his breath, double up, or drop to his knees. The blow is struck just under the breast bone or sternum.

14. OUTER EDGE OF FIST—JAW

The outer edge of the fist (little finger side) is protected more than the knuckles. An angle blow downward to the jaw separates the jawbones and breaks the jaw easily. It may be delivered from many positions. Hit lightly in practice.

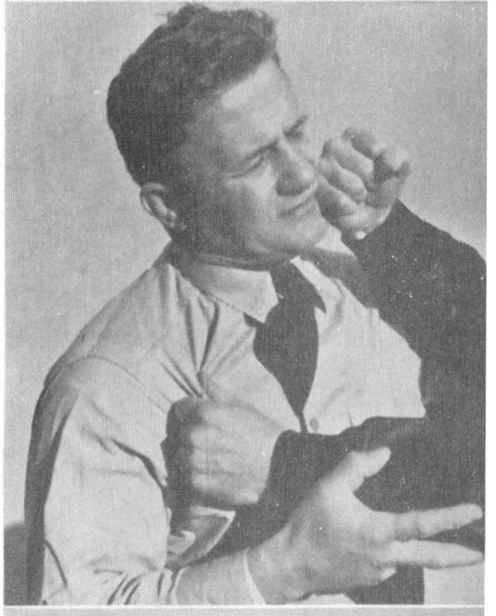

15. OUTER EDGE OF FIST—NOSE

This same blow may be delivered on the bridge of the nose. It is painful and causes eyes to water profusely.

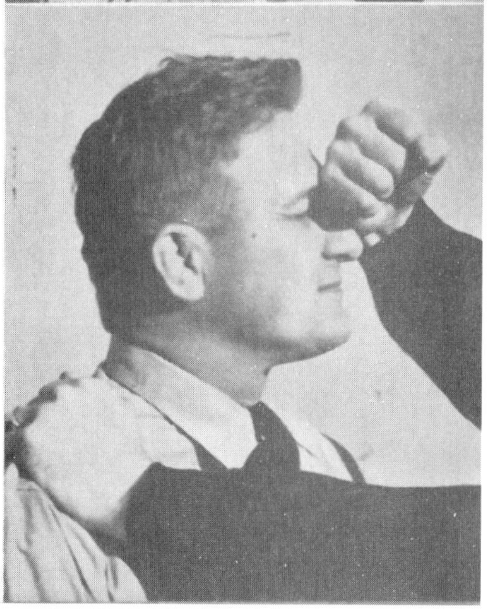

16. OUTER EDGE OF FIST—SHORT RIBS OR LIVER

Using the outer edge of the fist blow to the short ribs or over the liver is almost, if not as effective as the fist directly into the solar plexus. This is especially true if the opponent is struck unexpectedly. This blow may be delivered from the *front* or *rear*. If delivered over the short ribs from the right side it causes a severe pain to vibrate across the liver.

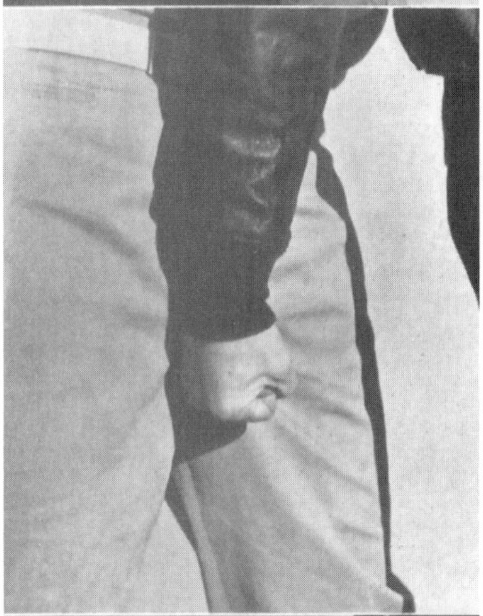

17. OUTER EDGE OF FIST—GROIN

This blow is struck when your opponent is behind you. Swing upward into the groin. This is dangerous and results in a quick finish of your adversary.

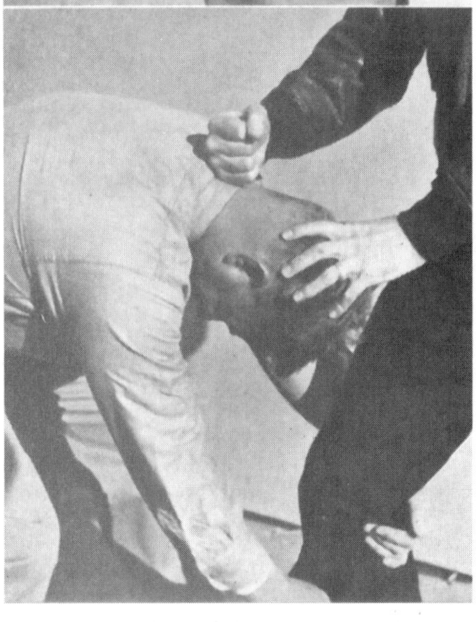

18. OUTER EDGE OF FIST—RABBIT PUNCH

This blow to the back of the neck may fracture the neck or severely injure your opponent. It may be used if your opponent attacks you with his head down or if you are behind him. Don't strike hard in practice.

19. OUTER EDGE OF FIST—KIDNEY PUNCH

Another effective use of the edge of fist blow is to the kidneys. It may be delivered from a variety of positions.

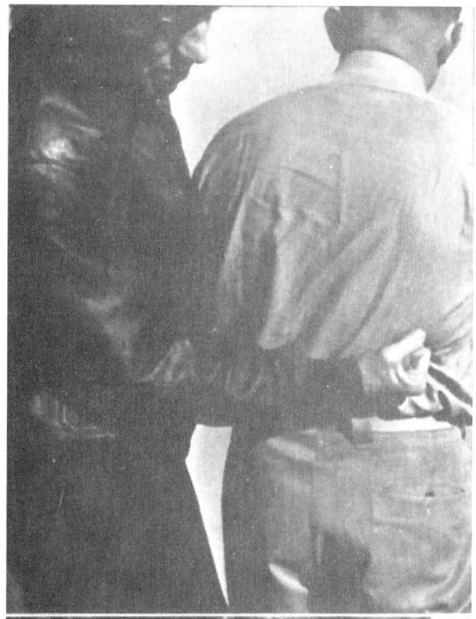

20. EDGE OF HAND

The knife edge of the hand, the edge on the little finger side, is well protected by flesh and skin. More damage will result from a blow with the edge of the hand than the closed fist because of its sharper edge, which gives a more penetrating blow over a smaller area. In practice, be careful not to strike too hard or sharply. The position of the hand, up or down, depends on the angle at which the blow is delivered, whether it be a direct swing or across the front of the body.

21. EDGE OF HAND—BRIDGE OF NOSE

A blow across the bridge of the nose will break this fragile bone tissue and bother vision for a short time at least.

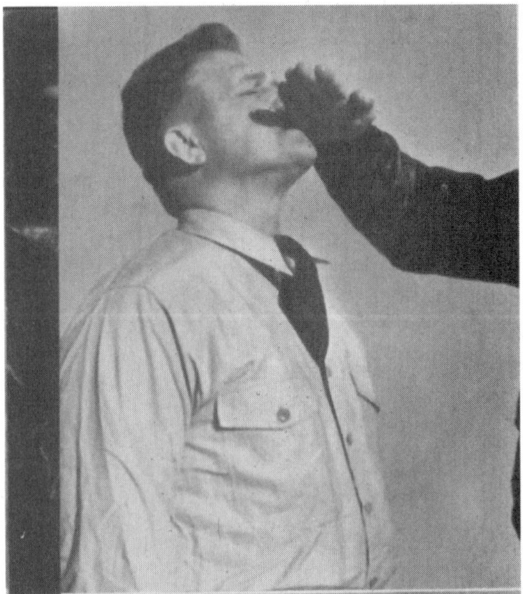

22. EDGE OF HAND—UPPER LIP

One of the most vulnerable spots on the face is that area just below the nose where the nose cartilage joins the bone. The nerves are close to the skin at this point. A blow struck slightly upward on the upper lip will have the effect you wish to create.

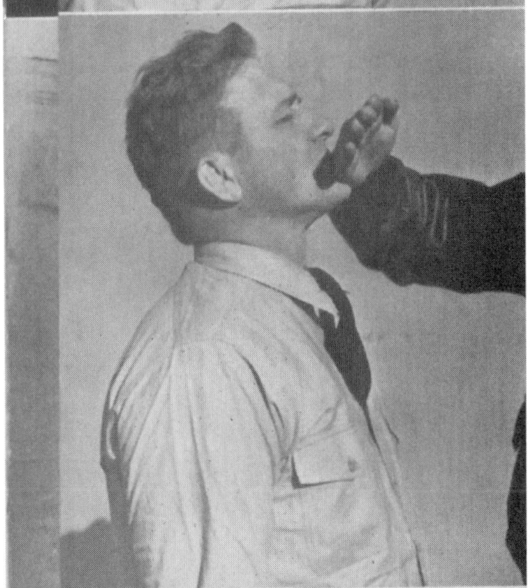

23. EDGE OF HAND—CHIN

A blow slightly downward on the point of the chin is painful and injures the lower jaw.

24. EDGE OF HAND—SIDE OF JAW

This blow has the same effect as the edge of fist strike to the side of jaw, but may be more severe.

25. EDGE OF HAND—SIDE OF NECK

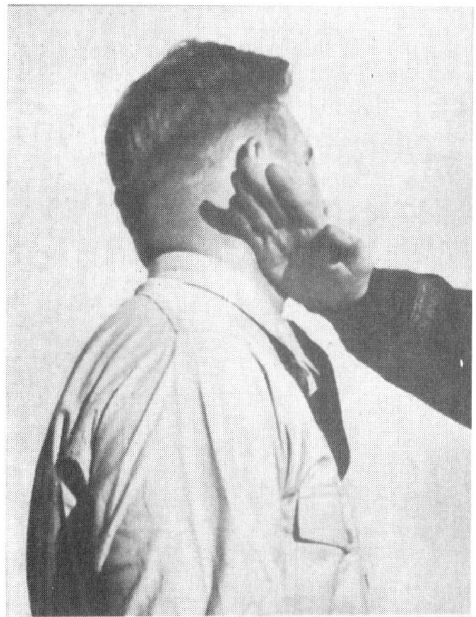

Deliver this blow to the side of the neck, below and slightly to the front of the ear. A sharp striking blow will shock the jugular vein, the carotid artery, and the vagus nerve causing a quick knockout. A strike to the temple can also be accomplished when you are in position to deliver the above described blow.

26. EDGE OF HAND—COLLAR BONE

Coming straight down on the collar bone with any great force will fracture the collar bone and quickly incapacitate your opponent. A blow delivered from this position to neck at shoulder will bring a man to his knees.

27. EDGE OF HAND—SOLAR PLEXUS

A blow just below the breast bone or sternum with the knife edge of the hand can be more effective than a blow with the fist due to the sharpness of this side of the hand with its penetrating effect.

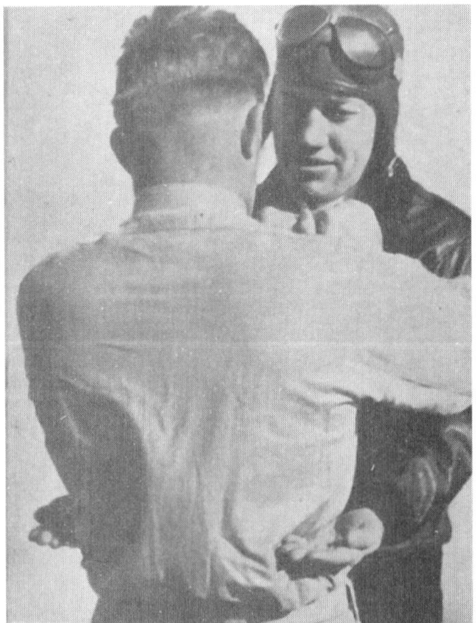

28. EDGE OF HAND—SHORT RIBS

One or both hands—can deliver this blow to the short or floating ribs on one or both sides, front or rear. The bodily shock is terrific, particularly on your adversary's right side just over the liver.

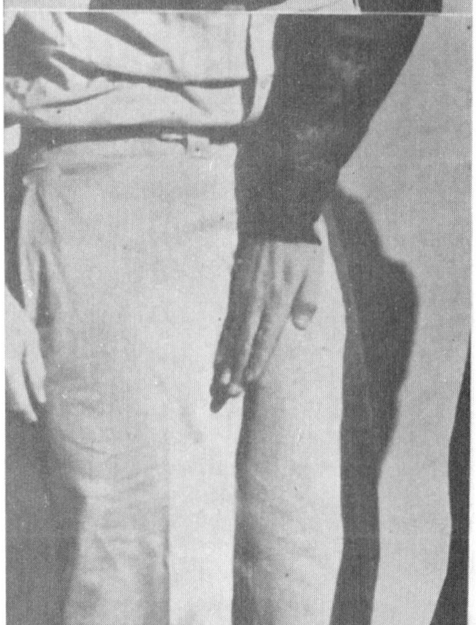

29. EDGE OF HAND—GROIN

A blow to the groin is always effective.

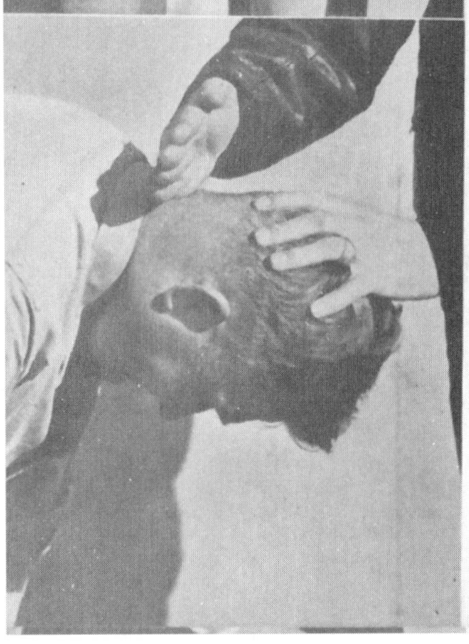

30. EDGE OF HAND—NAPE OF NECK

The "rabbit punch" can kill a man. This maneuver will always work well whenever the attacker comes at you with his head down. This blow can break the neck.

31. EDGE OF HAND—KIDNEY

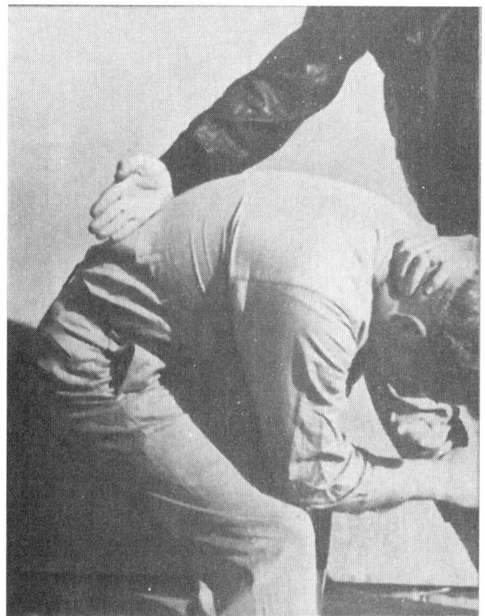

When your opponent's back is turned partly toward you, a shocking blow to the kidneys may be delivered. Such a blow may rupture a kidney and if not given immediate medical attention death will result. This blow is extremely painful.

32. HEEL OF HAND—CHIN

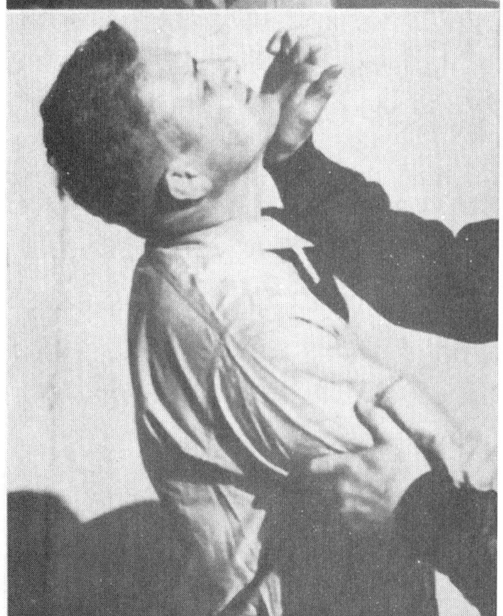

At comparatively dose quarters, the heel of the hand can deliver some devastating blows as this upward strike to the chin indicates.

33. HEEL OF HAND—NOSE

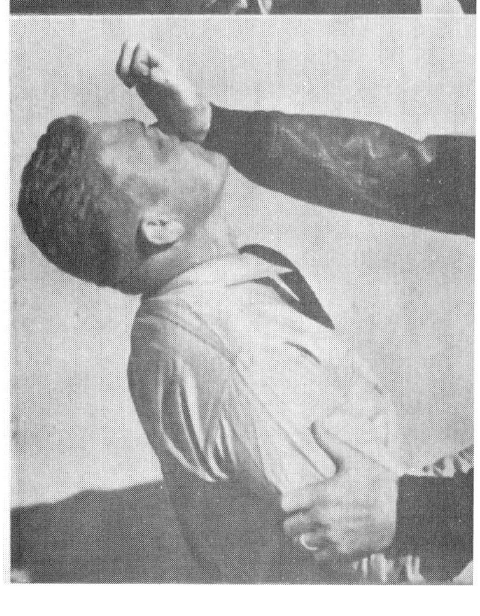

This same blow to the nose, upward, will loosen any grasp your opponent may have on you. Excellent for close quarters. Also good for breaking apart persons locked in combat.

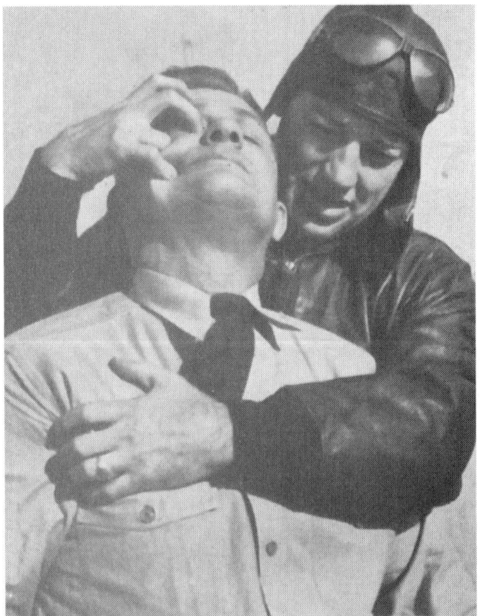

34. FINGER POKE—EYES

It need not be explained what effect the finger poke in the eye will have on a man. Gouging is not fair fighting, but it comes in handy if your life depends on it.

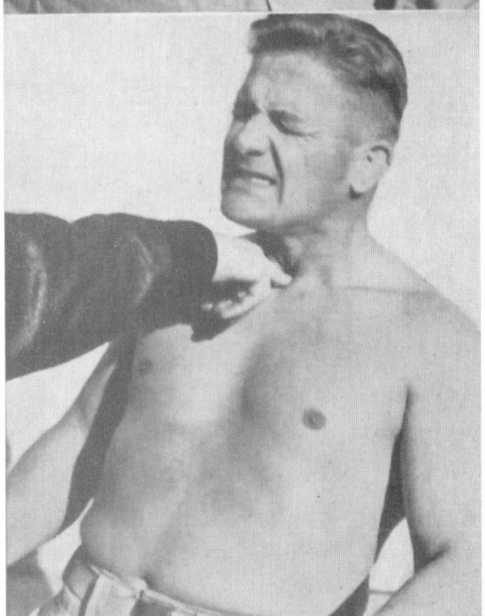

35. FINGER POKE—BASE OF THROAT

If the finger is pushed into the hollow space just above the breast bone or sternum enough pain results to cause your attacker to release his hold.

36. GRASPING HAND—JUGULAR VEIN AND CAROTID ARTERY

Using the fingers to grasp the side of the throat firmly, tear this area away as far as you can. By the same token, your teeth imbedded in this region can cause a great deal of damage, especially when you get behind your adversary. Don't be afraid to bite at any part of your assailant's anatomy if you can reap any beneficial results from so doing.

37. GRASPING HAND—THROAT

The hand and fingers securely grasping the Adam's apple or larynx can cause serious injury. The pressure may be applied by squeezing or pulling outward.

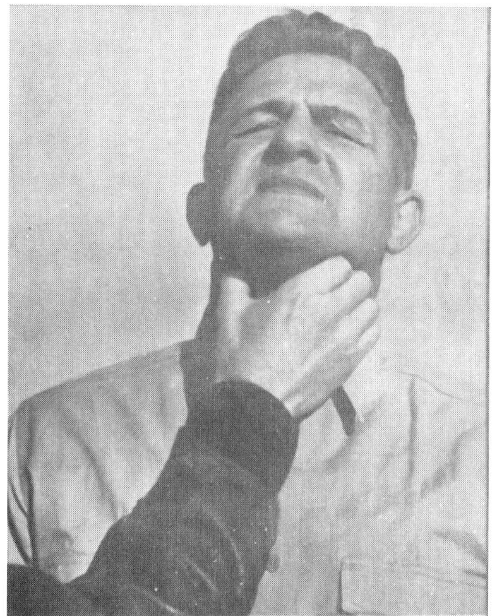

38. GRASPING HAND—LIPS

This pressure can be applied to the upper lip at the base of the nose or to the lower lip. It is very painful. Try it and see how it feels.

39. GRASPING HAND—GROIN

May be obtained from a variety of positions. Obtain a firm grasp and jerk downward for serious results. This is extremely dangerous.

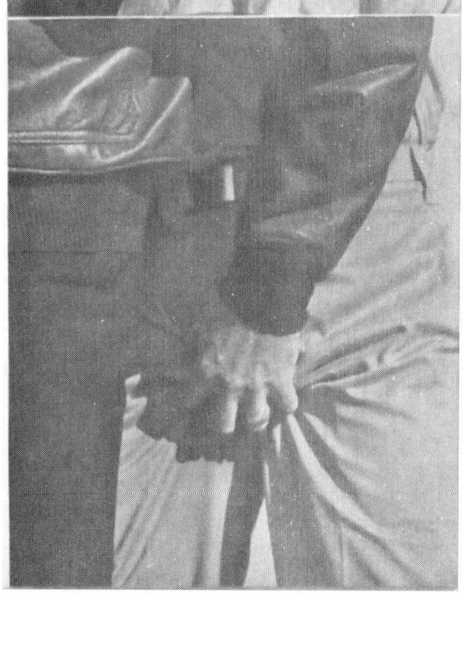

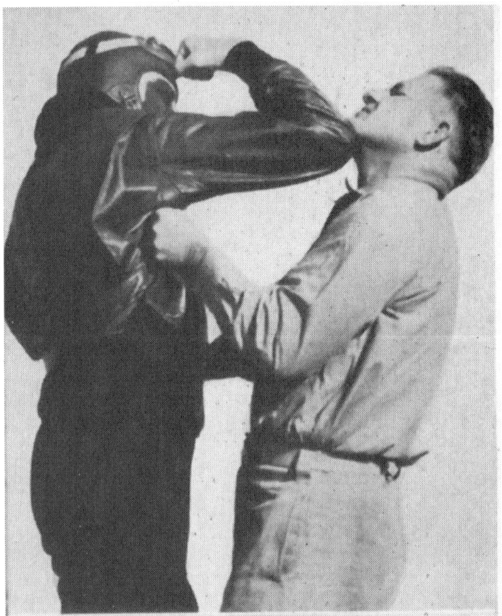

C. Elbow as Weapon

40. FRONTAL ATTACK—UP TO CHIN

When in close quarters bring the elbow up to the chin. A great deal of power can be put behind this blow, making it a great subduer. This maneuver works exceedingly well when you are in a crowded position or if your adversary is taller than you.

41. FRONTAL ATTACK—DOWN TO CHIN

In this instance the elbow is brought down to the chin for the desired results. Works best on a shorter opponent or when he is in crouched position.

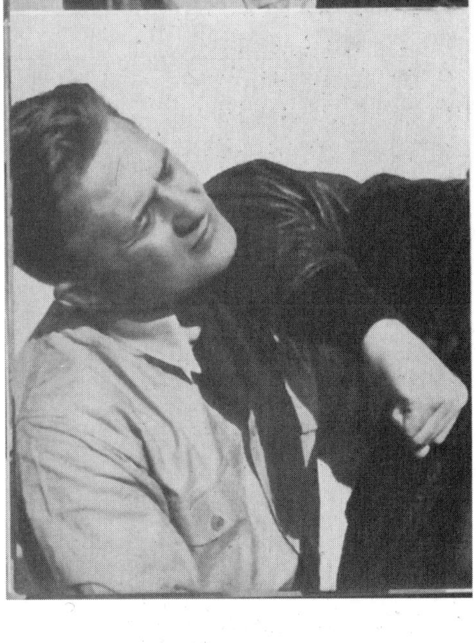

42. FRONTAL ATTACK—FRONT LATERAL TO CHIN

The elbow blow comes from the right or left and strikes against the side of the jaw at the chin. The blow is struck with a forward motion.

43. FRONTAL ATTACK—BACK LATERAL TO CHIN

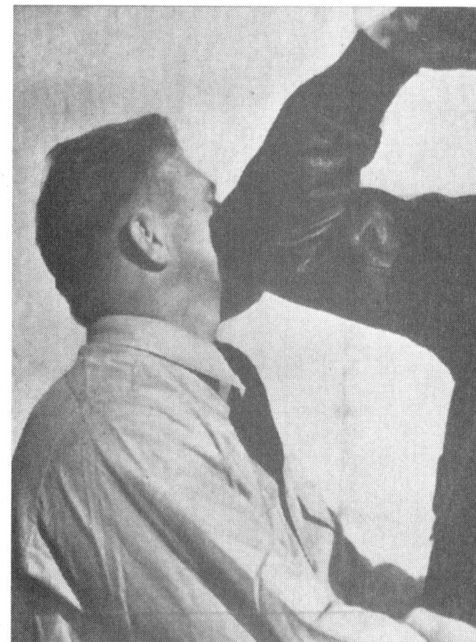

Actually this blow is the result of the elbow returning from a strike to the chin (as shown in illustration 41), but delivered to the opposite side of the jaw at the chin. This blow and the one in illustration 41 are very effective for quick knockouts at dose quarters.

44. FRONTAL ATTACK—DOWN TO COLLAR BONE

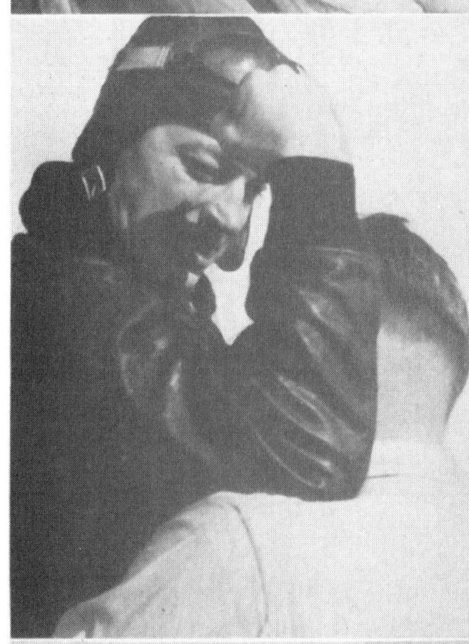

Such a blow delivered down against the collar bone win break it. It is not necessary to use every ounce of strength to accomplish this result. Works most effectively when an opponent is shorter or in a crouched position. In all the elbow blows never strike with the arm extended; be sure your elbow is bent and the muscles tensed against the shock.

45. REAR ATTACK—SNAP BACK TO FACE

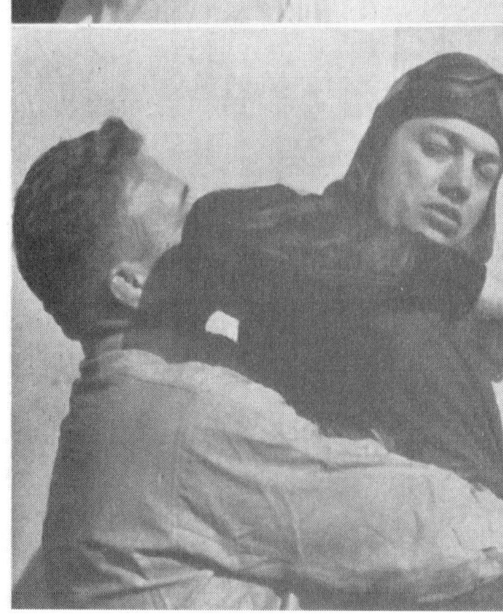

Whenever an assailant is behind at close quarters or is grasping you with a body lock so that your arms are free, quickly turn and strike with your bent elbow to his face. Do this with quick jerks and turns to either side. Don't go too high. You might injure your elbow on the hard parts of your adversary's head. Aim for the face, temples, and neck.

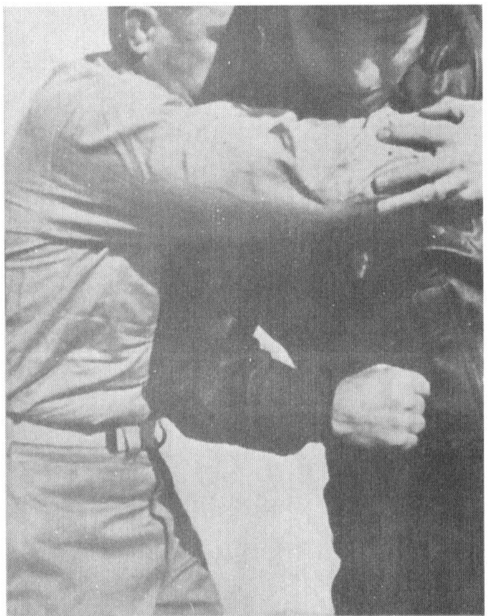

46. REAR ATTACK—SNAP BACK TO SOLAR PLEXUS OR RIBS

Again, if your adversary is behind, snap your elbow back into his ribs or solar plexus. If he has seized you about the waist in a body lock, this maneuver will not work. In such instances you should strike to the face (illustration 45). If your adversary has pinioned your arms to your sides, raise the arms, squatting slightly to give you "elbow room," and strike back into the short ribs of solar plexus.

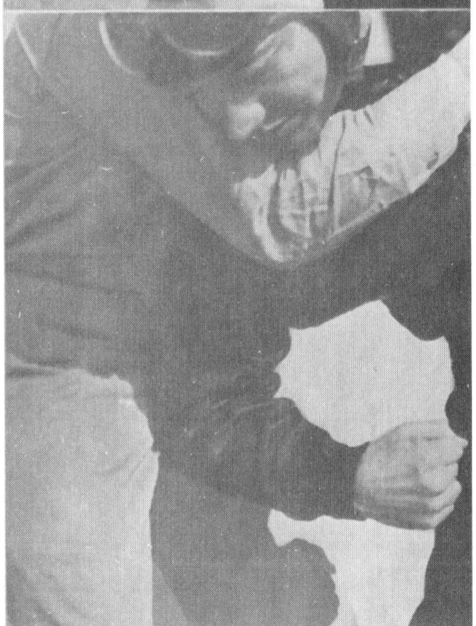

47. REAR ATTACK—SNAP BACK TO GROIN

By dropping or squatting lower, the elbow can be directed at the groin.

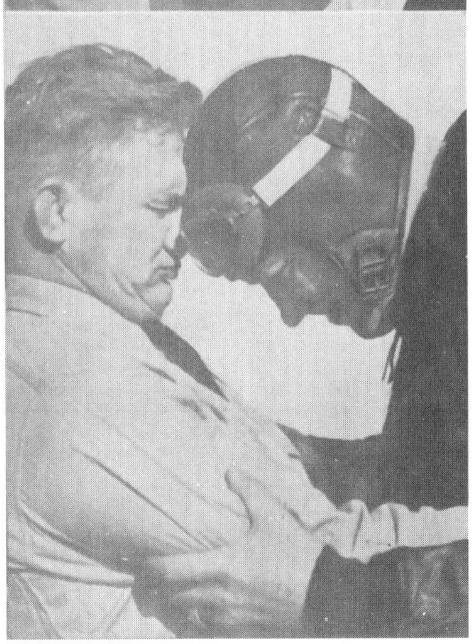

D. Head as Weapon

48. FRONTAL TO FACE

If body locked from the front or in close quarters with an enemy, bobbing the head (forehead or top-first) into the face, from chin to nose, will effectively injure him. It need not be stressed bow much more effective this maneuver would be if you were wearing a steel helmet. Don't try this on your friends—it's dangerous.

49. REAR TO FACE

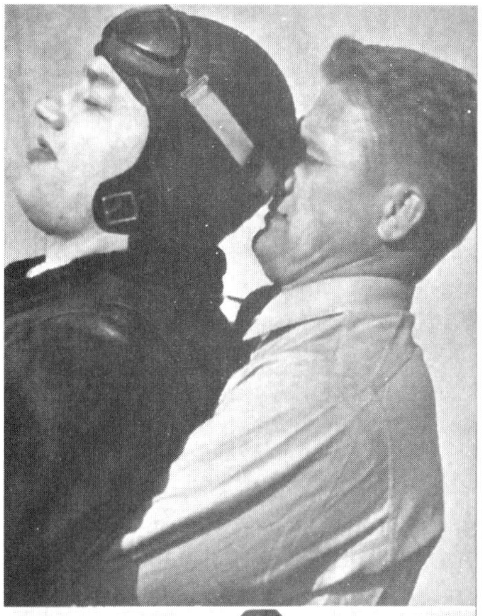

In this maneuver the back of the head is the weapon. Drive the head back into your assailant's face. This is usually effective for breaking body locks from the rear.

E. Knee as Weapon

50. KNEELIFT—FACE

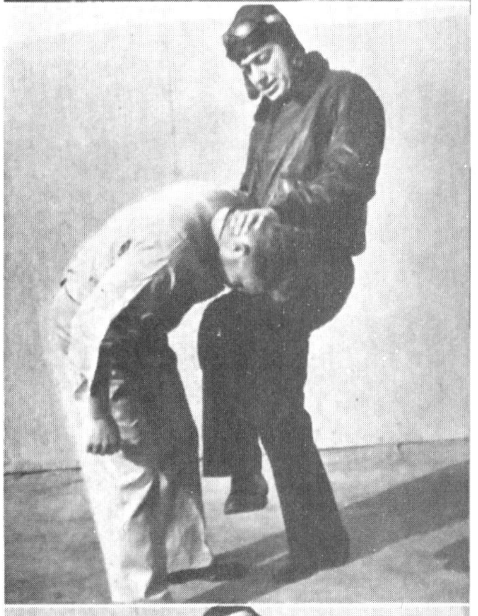

The kneelift or raising the knee sharply is an excellent weapon. In this instance it is used in the face of an opponent. It can be used if he attacks in a crouched position, tries to tackle, or has been forced to bend over. It can be used as a subduer any time you can bring your opponent's face within striking range. You will discover the effectiveness of its use after a study of the many combinations illustrated later in this manual.

51. KNEELIFT—NECK

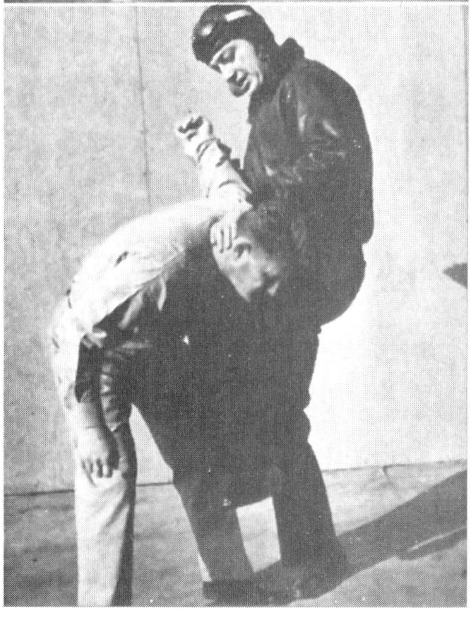

In order to snap the knee quickly into the neck, the situation is practically the same as illustration 50.

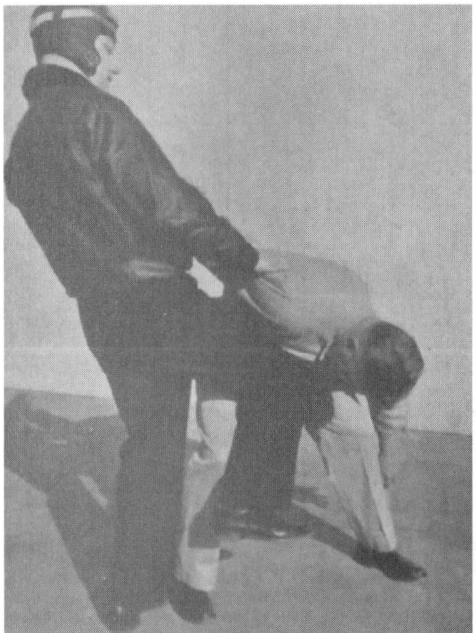

52. KNEELIFT—CHEST

Whenever the face and neck are not within range, a quick snap into the chest or ribs slows up your opponent.

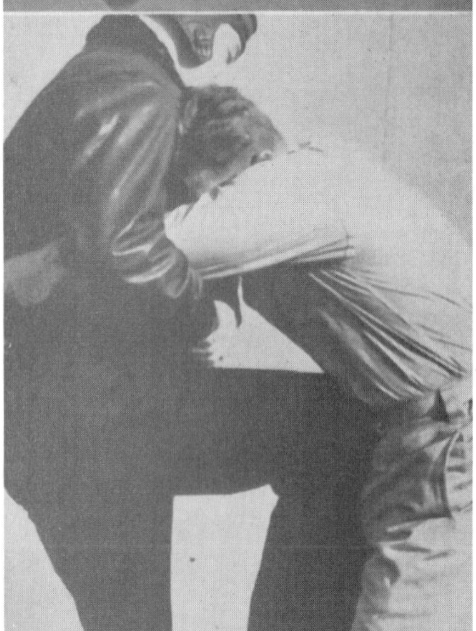

53. KNEELIFT—STOMACH

The kneelift to the stomach or solar plexus is as good as a fist blow to the same area.

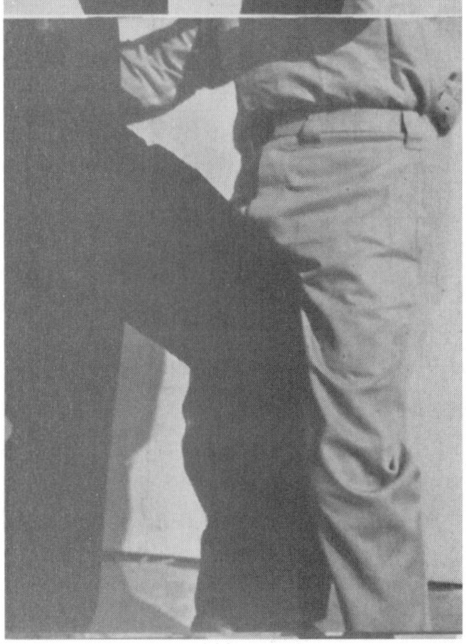

54. KNEELIFT—GROIN

If an opponent is facing you at close quarters, this blow to the groin will conclude the affray immediately. Don't be afraid to use it under dire circumstances, but be careful when you practice. Get in the habit of lifting the knee.

55. KNEELIFT—BACK (SPINE OR KIDNEYS)

If an adversary has his back to you or is bent over backward, a knee lift to the spine or kidneys will result in a great deal of damage.

56. KNEE DROP—HEAD

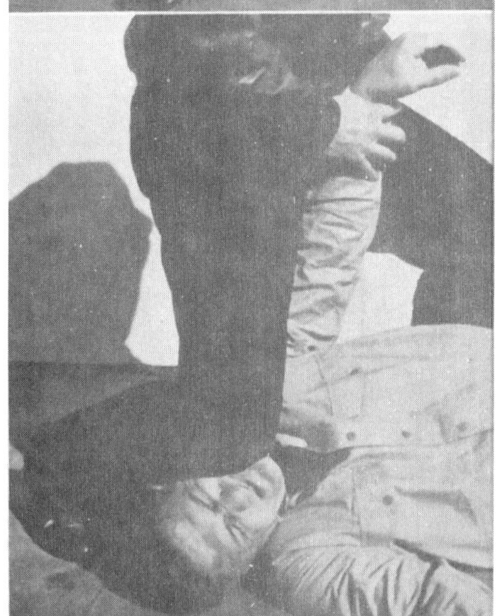

The knee drop is accomplished when you have your assailant on the deck. The bent knee drops to his body, with the entire weight of your body behind it Dropping this weight on the vulnerable parts of an adversary's body with the sharp knee concentrates a great strain to a small area. Almost always broken bones result. Here the knee drop is directed at the head. If the head is against the deck, the weight behind the knee has a crushing effect on the skull and may result in a fracture. Watch out in practice—drop very lightly.

57. KNEE DROP—NECK

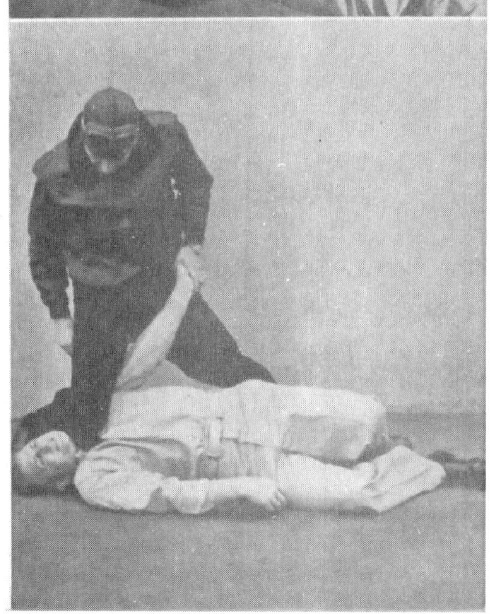

The same maneuver to the neck will fracture the neck vertebrae.

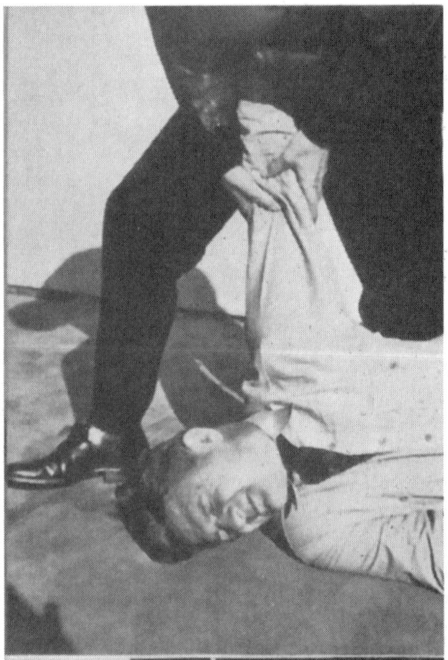

58. KNEE DROP—RIBS

A knee drop to the ribs will break them and drive broken bones into the lungs.

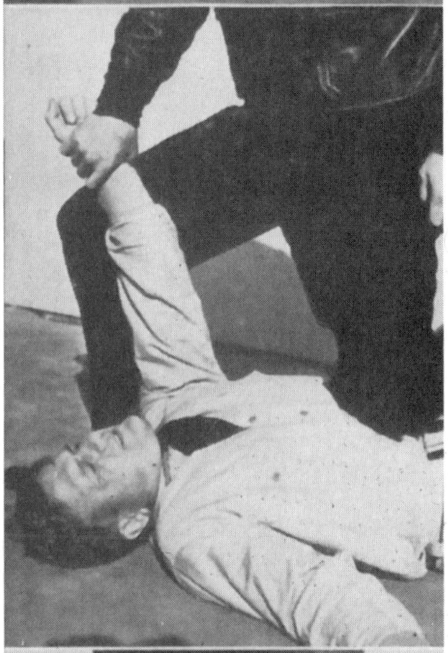

59. KNEE DROP—STOMACH

The knee into the stomach or solar plexus may injure an adversary seriously.

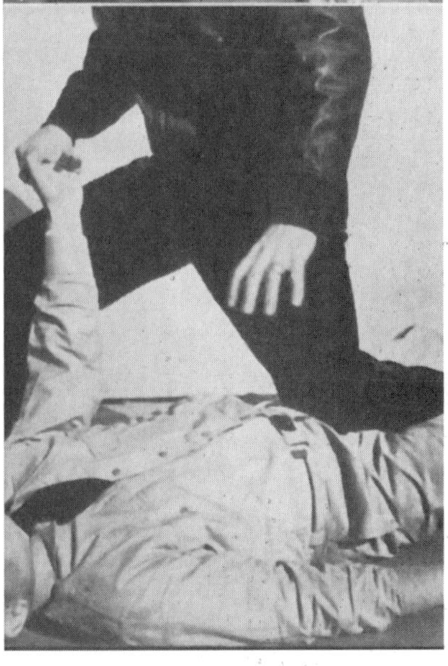

60. KNEE DROP—GROIN

This blow has the effect of a kneelift to the same area, but is more severe. The battle should conclude at this point.

61. KNEE DROP—BACK (SPINE OR KIDNEYS)

If your opponent is in this position, you can break his back or rupture his kidneys by driving your knee down with your weight behind it.

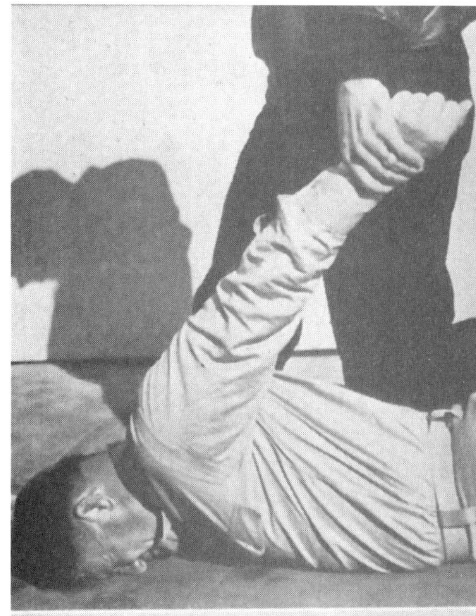

F. Foot as Weapon

62. TOE—SHIN

Of course, these maneuvers are effective only if shoes are worn, and the heavier the shoe the more severe the consequences. The toe kick may be delivered whenever an opponent comes within range. This kick to the shin is extremely painful and is effective in disconcerting an adversary. In other words, it is not a killing blow, but it does hamper and distract the foe from his original thoughts.

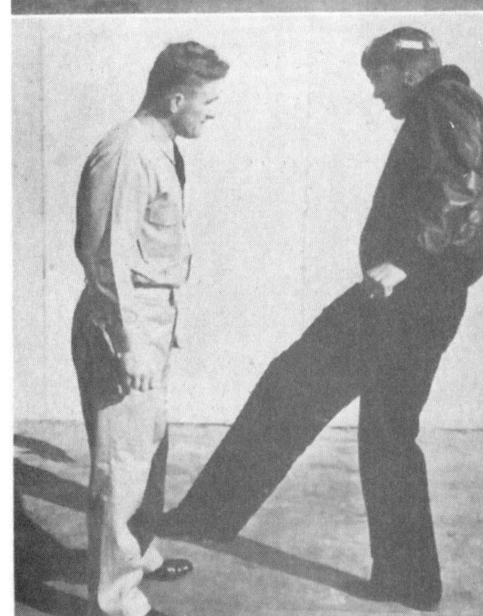

63. TOE—KNEECAP

A toe kick to the kneecap is always an effective means of attack. It can be delivered when one is within range and from several positions. The kneecap or patella is a round floating bone, attached at the top and bottom by tendons, which protects the knee joint from injury. Any severe blow to this area seriously affects a person's ability to stand up or walk.

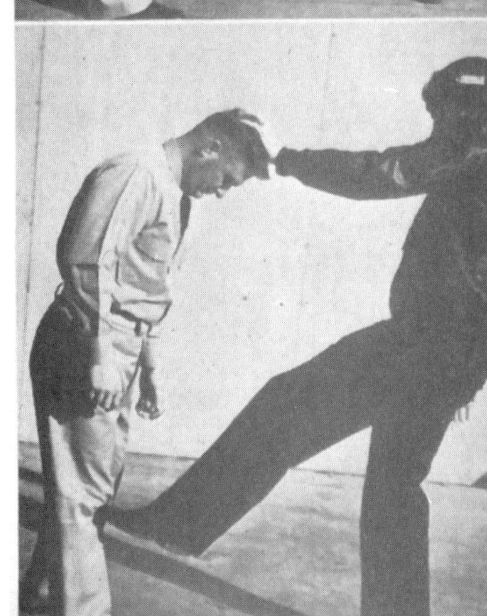

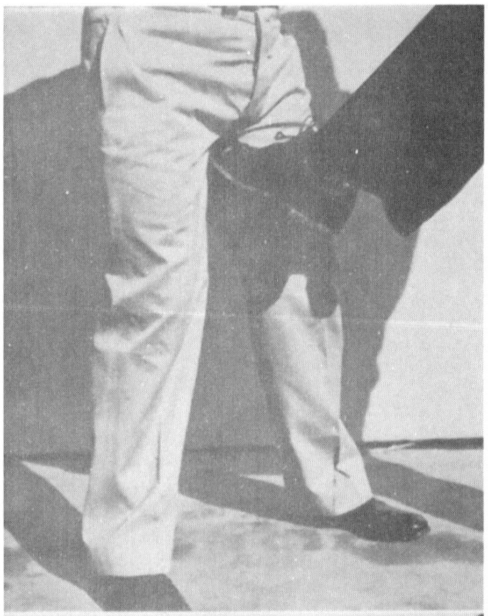

64. TOE—GROIN

The toe of the shoe brought up into the groin will injure an opponent seriously. It is always best to aim your blow in toward his crotch. If the toe misses, it will pass upward between his legs and you will catch him in the groin with your shin.

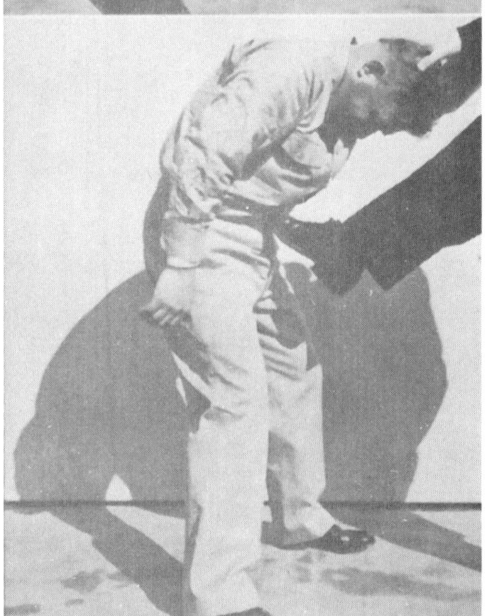

65. TOE—STOMACH

This kick may be delivered whenever an adversary is crouched or slightly bent over.

66. TOE—CHEST

A solid toe kick to the ribs or chest will break the ribs and usually incapacitate your assailant.

67. TOE—HEAD

Kick to the face or head when they are in range. In French savate, this kick to the face is often delivered to an upright opponent, the foot being brought straight up. It can be done easily with a little practice and the feet can become as effective weapons as the hands

68. TOE—HOCK OF KNEE

This kick delivered from the rear to the hock or back of the knee with any amount of force penetrates the flesh to the nerves, paralyzing the lower limb. It is extremely painful.

69. TOE—COCCYX (TAIL BONE)

A toe kick to the base of the spine not only is painful, but may break the bone. It is delivered from the rear, whether your opponent is standing or lying on the deck.

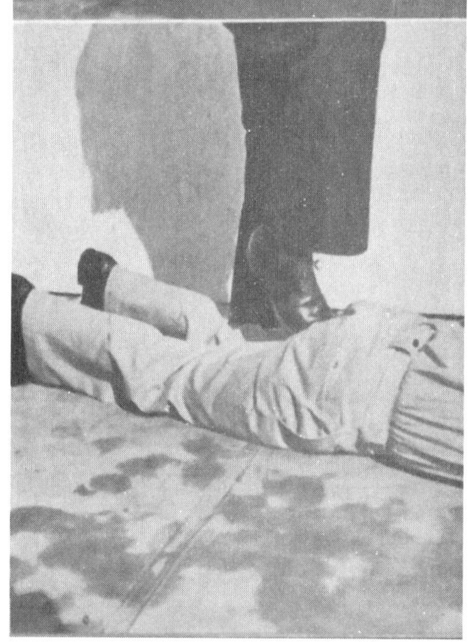

70. TOE—SPINE

Unusually effective to subdue a foe quickly. Kicks to the spine are always dangerous and may cause serious injury.

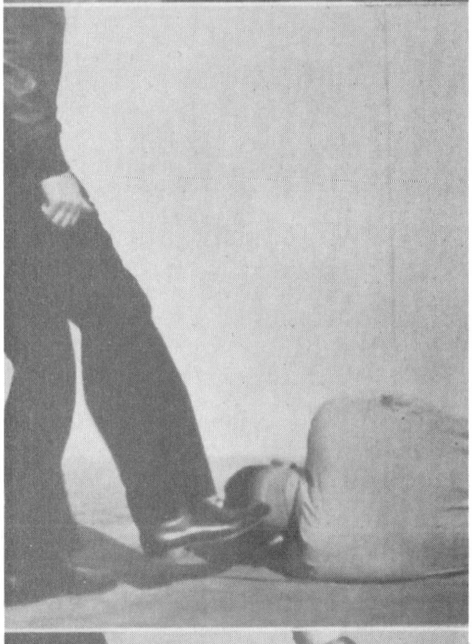

71. TOE—NECK (REAR)

The same kick to the base of the neck eliminates an opponent quickly.

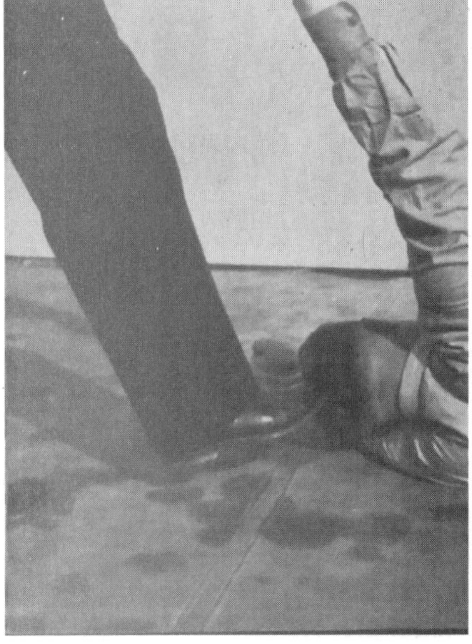

72. TOE—HEAD (REAR)

A toe kick with a hard shoe to the head easily may cause a fractured skull or death. It is a quick means of subduing an adversary when he is on the deck.

73. HEEL—TO INSTEP (FRONTAL)

An adversary's feet are generally within range of a heel blow downward to the instep. This maneuver is applicable from many positions. Illustration 73 shows a frontal attack with the heel driven sharply down.

74. HEEL—TO INSTEP (REAR)

The heel may be brought down with force to the instep if your assailant attacks from the rear. The blow may not be sufficient to dislodge any grasp he may have, but it will cause him to flinch or shift his position, all that is needed to work many of the maneuvers listed in the manual.

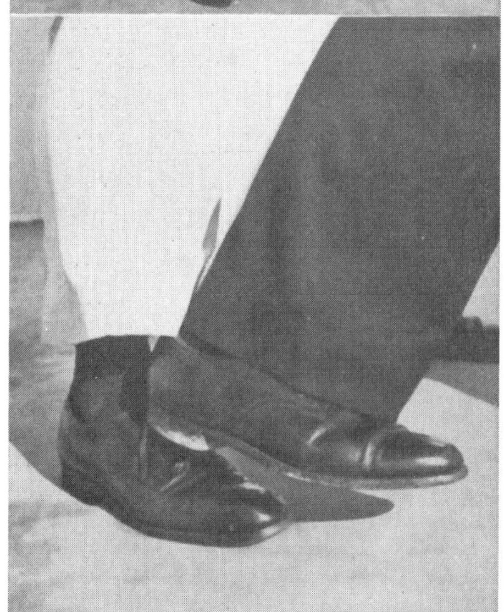

75. HEEL—SHIN (REAR)

A heel brought back sharply against the shin when an assailant is behind will have the same effect as the blow to the instep. A blow to the shin, delivered from the side with the side of the shoe is just as effective.

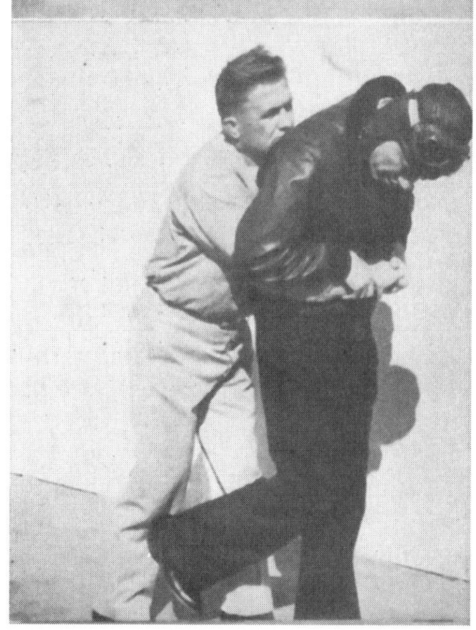

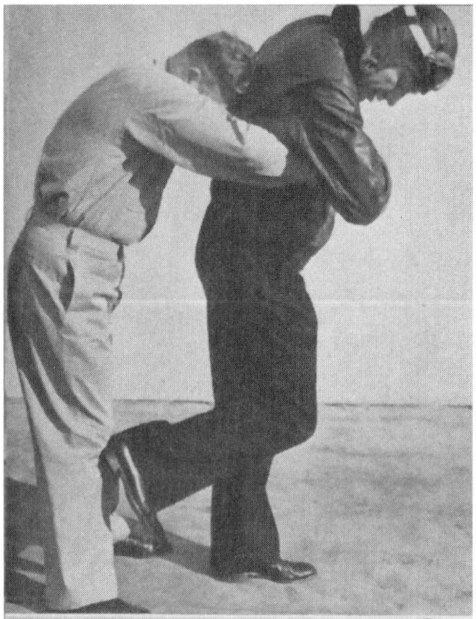

76. HEEL—KNEECAP (REAR)

A blow against the kneecap is just as effective as the toe kick (illustration 62). Lunge forward slightly and snap the heel sharply upward.

77. HEEL—GROIN (REAR)

This blow is delivered after a quick lunge forward in order to get room to work. The heel is snapped upward sharply. Your adversary may have you clinched in any one of a variety of holds, but the kick can be accomplished with a little quickness on your part.

78. HEEL—HOCK OF KNEE

The heel is quickly jabbed into the hock or back of the knee with force. This maneuver is very effective for use if a prisoner attempts to run or escape when you are behind him and within range. Kick or jump quickly to the hock of the knee and bring him to the deck. A terrific strain is placed on the knee.

79. HEEL—SMALL OF BACK

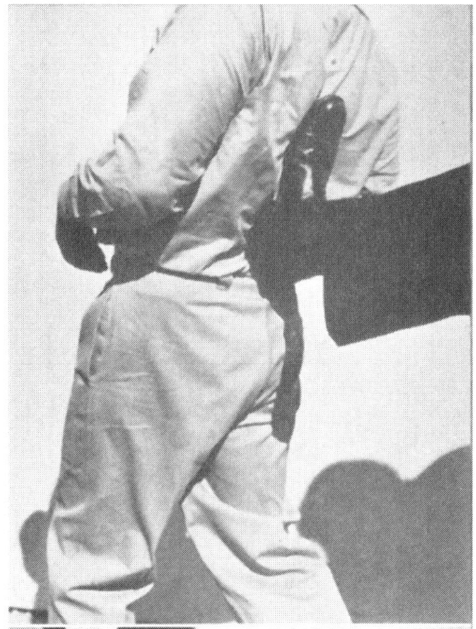

A quick kick with the heel into the small of an opponent's back is a very effective maneuver, especially if the recipient is not expecting the blow. The entire strength of leg can be put behind the blow.

80. HEEL—BACK OR NECK (ON DECK)

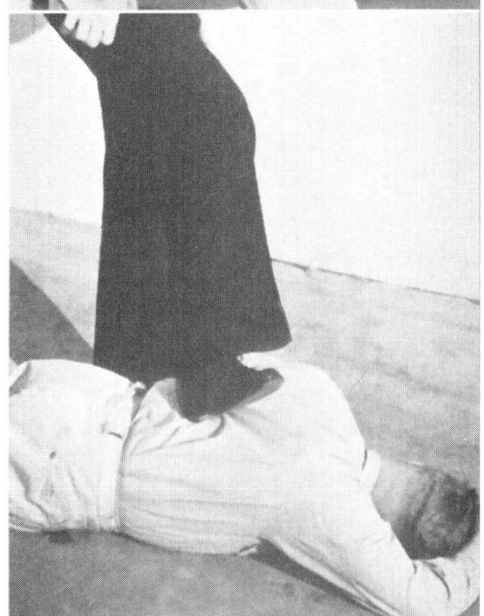

At any time, when you have an adversary on the deck face down, the heel brought heavily or sharply down into the back, spinal column, kidneys, or neck will cause a serious injury or eliminate your victim altogether. You will find the use of the feet, especially the heels, essential to the completion of many techniques listed and discussed by this manual.

81. HEEL—RIBS (ON DECK)

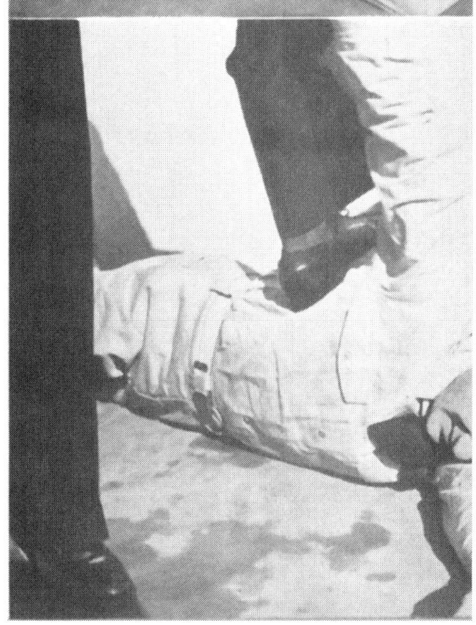

When an assailant is in position, deliver a sharp downward blow to the ribs with the heel.

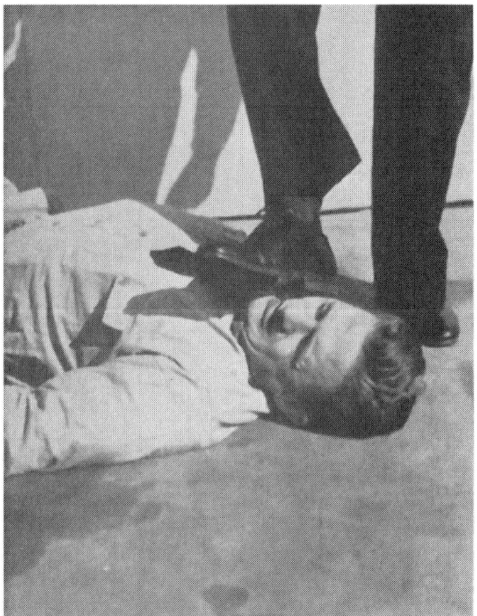

82.. HEEL—FACE (ON DECK)

A heel kick to the face is probably the quickest way of changing one's complexion. It causes serious injury or even death.

G. Nerve Shock as Weapon

(Showing a variety, but not all methods of attack. See charts—Chapter IV.)

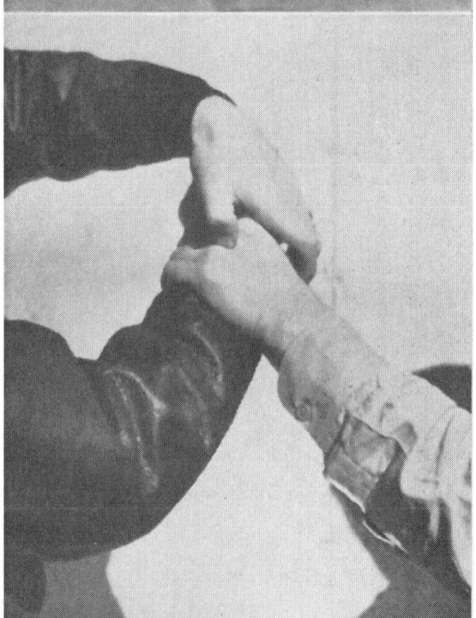

83. NERVE SHOCK—SECOND AND THIRD KNUCKLE—BACK OF HAND, THUMB PRESSURE

Pressing the thumb or thumb nail into the back of band between the 2nd and 3rd fingers stimulates the nerve in this area, causing an adversary to flinch perceptibly or to release his grasp on your body or clothing. Note location of the thumb pressure—it is down between the bones of the fingers of hand and against the ring finger knuckle.

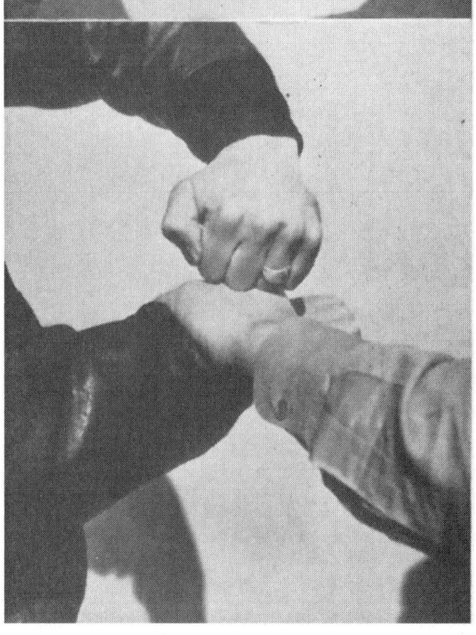

84. NERVE SHOCK—KNUCKLE JAB TO BACK OF HAND

In order to make an adversary release a grasp on your clothing or body, strike across the back of his hand very sharply with the knuckles. The blow does not cause any injury, but is painful, disconcerting, and generally makes an assailant release his grip.

85. NERVE SHOCK—KNUCKLE ON FORE-FINGER NEXT TO THUMB

Any pressure against the forefinger, as shown in this illustration, is painful and tends to make an Adversary flinch or release his grip.

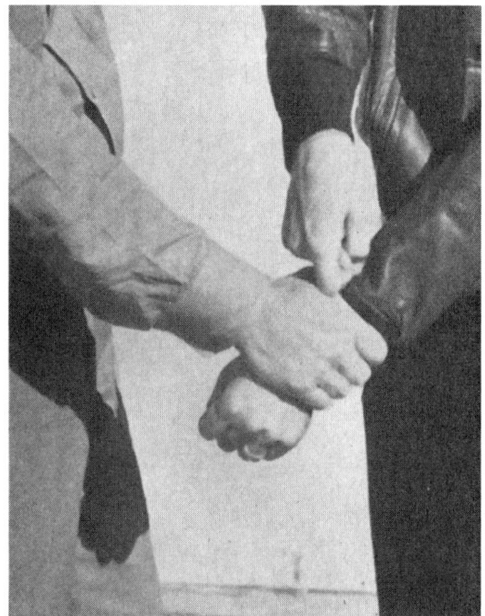

86. NERVE SHOCK—WRIST

Digging the fingernails into the inside of the wrist and rolling them back and forth causes them to pass over the nerves in this area, shocking the adversary momentarily. This allows you to act as you see fit.

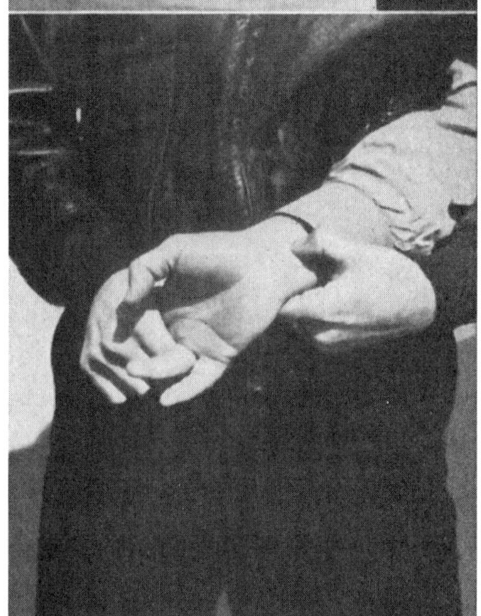

87. NERVE SHOCK—ELBOW

Just above the elbow on the inside of the arm is a very sensitive nerve. This nerve, when pressed by the fingers or thumb, affects the entire arm. You have, on occasion, bumped your "funny bone" and know how painful it can be and how useless your arm is for a short while. The pressure on this nerve creates the same effect.

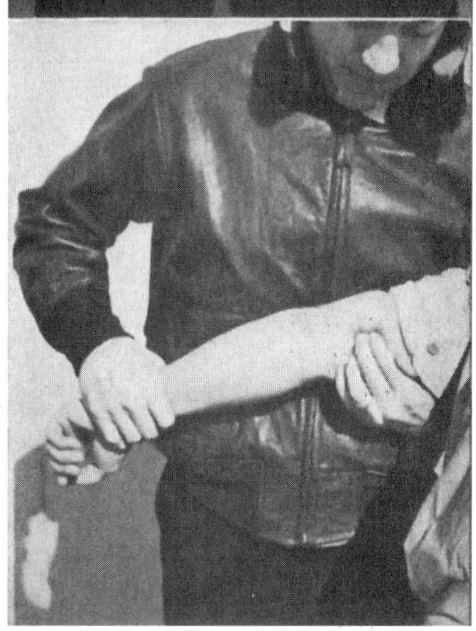

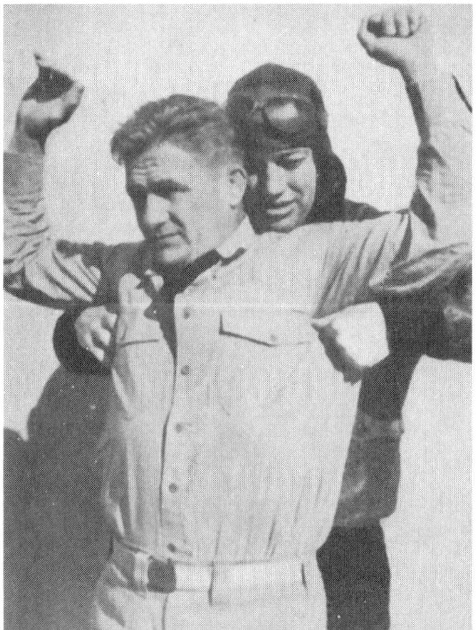

88. NERVE SHOCK—ARMPIT

Deep in the armpits are located nerves, which, when the arms are raised, are exposed to finger as a pressure. Any pressure inward and upward has a tremendous effect on the victim. This pressure is excellent for use when one does not wish to be pushed through a doorway or out of a vehicle. Of course, it is difficult to apply through clothing, such as coats and overcoats; therefore, reach around in front and pass your hands under the clothing at the chest. Now reach back to the armpits. It will be noted that this pressure is sometimes ineffective on fat individuals, due to a layer of fat covering the nerves; in such cases use the knuckles and dig in sharply and powerfully.

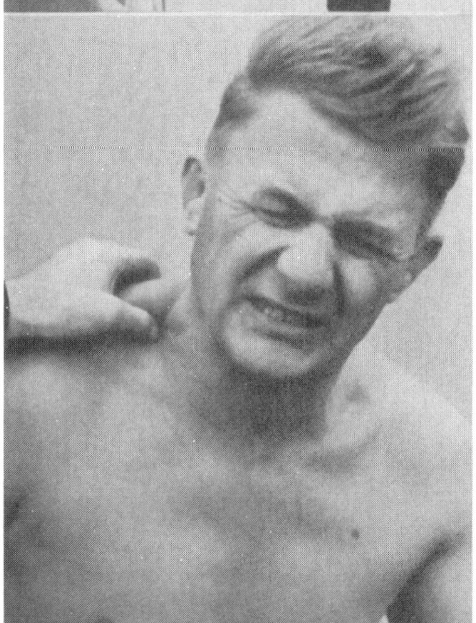

89. NERVE SHOCK—SHOULDER

Note how the muscle at the shoulder near the base of the neck is pinched. You don't grasp or pinch the entire muscle, only part of it. It is an effective means of releasing yourself from a critical situation.

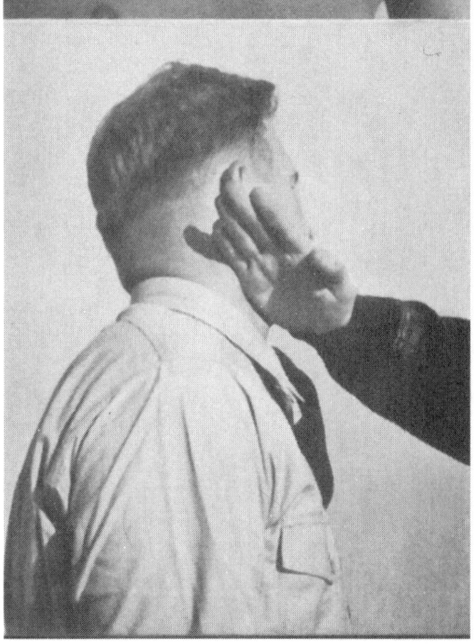

90. NERVE SHOCK—VAGUS NERVE

(Note the edge of hand strike in fig. 25.) This strike with the edge of the hand so shocks the nerve that a quick knockout is inevitable. This blow to the nerve is dangerous; it may result in death to a person suffering from heart trouble. Be extremely careful of its use in practice. A pressure on this nerve with the fingers or with the arms, as illustrated in 119, is also effective, as it causes the victim to lose consciousness.

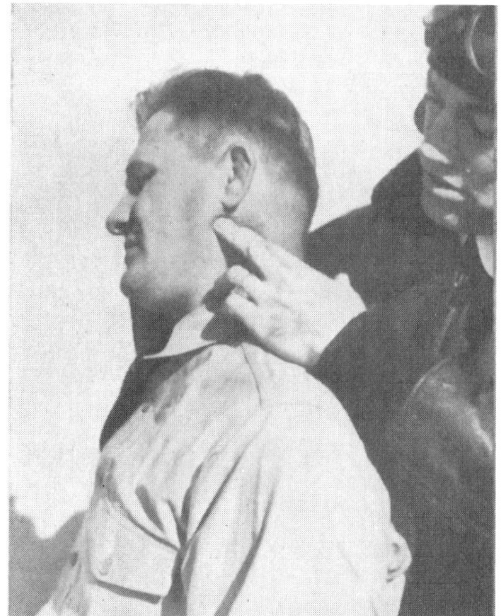

91. NERVE SHOCK—BELOW EAR

This pressure, shocking the nerves, is applied just under the ears behind the jaw. It is excellent for raising a person from a sitting position. It may be applied from either the front or rear.

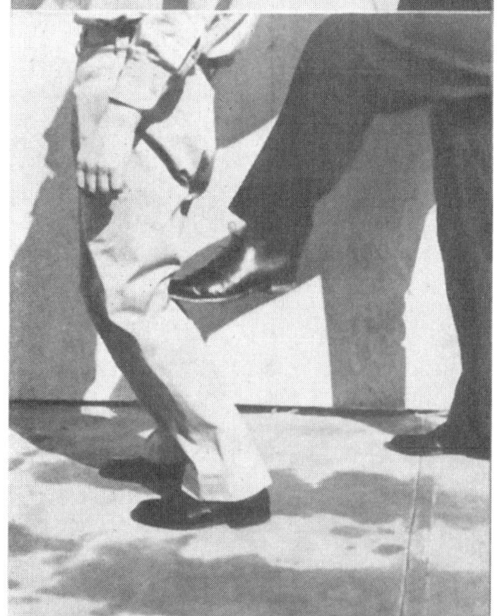

92. NERVE SHOCK—HOCK OF KNEE

This nerve shock may be applied with the toe (as in illustration 68) or by a severe blow with the knife edge of the hand or the outer edge of the fist. These nerves are deep seated, but can be contacted if the blow is sharp enough.

H. Fingerlocks

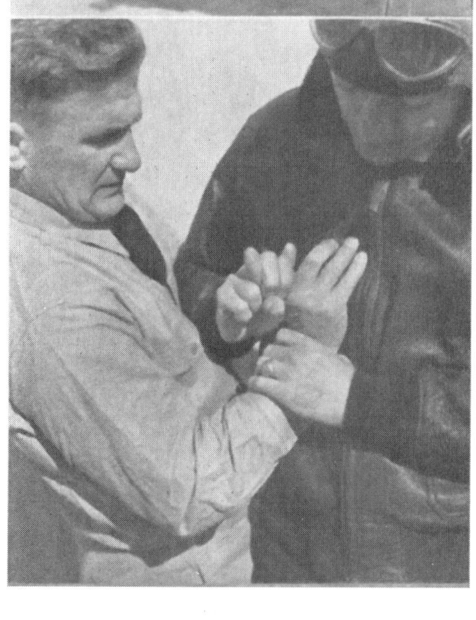

93. ONE FINGER SNAP

Two fingers are wrapped around adversary's one for power and leverage. If his palm is downward, the snap is up and back; if palm is up, snap down and back. Your other hand secures the wrist from the top or bottom, depending on the situation at the time. Pressure, if applied sharply and strongly, will break the finger. A quick steady pressure will bring your adversary to his knees. The maneuver may be used in a variety of situations, whether adversary is grasping your clothing, hair, or body in any fashion from the front.

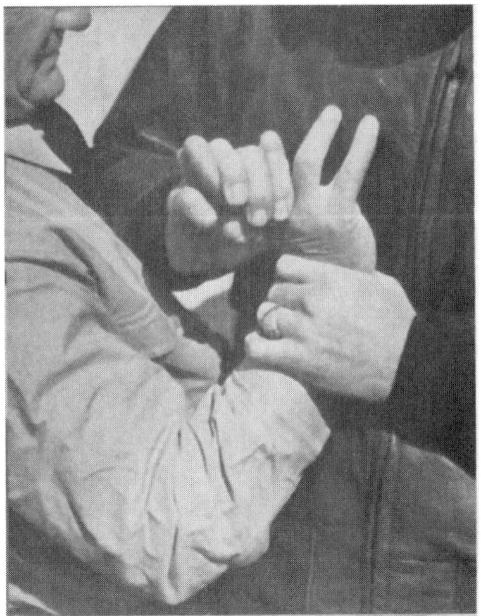

94. TWO FINGER SNAP

This maneuver is the same as in illustration 93 except that two fingers are grasped by your three fingers or more. In these maneuvers, it is not necessary to grasp the fingers shown. It will work on any finger or fingers.

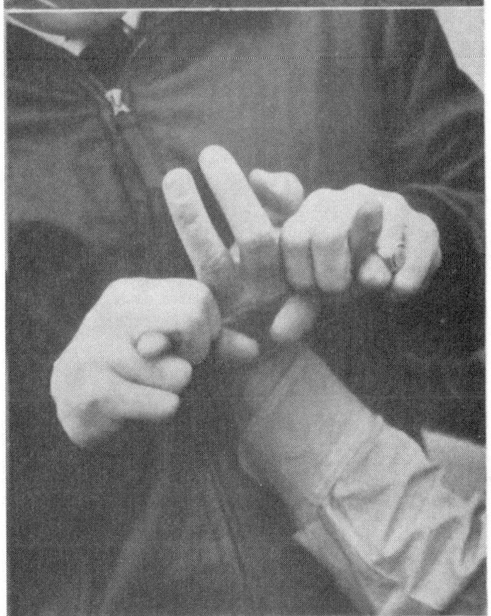

95. FINGER SPREAD

For most effective results, bear sharply backward and outward with each hand to get the spreading movement. If the palm is up, the pressure is downward and outward. A quick snap will break the fingers or bring assailant to his knees.

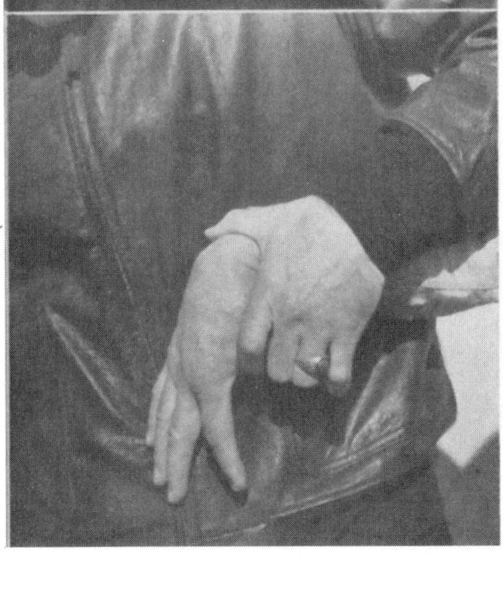

96. THUMBLOCK

Note how the fingers pass over the thumb and the inside of the wrist. The thumb is against the back of the wrist. This maneuver is not particularly effective against persons with limber wrists or double jointed thumbs; in such cases use with a spread to another finger.

97. FINGERS (REAR)

When secured by a body lock from the rear, the grasp may be broken by bending any finger or fingers back against the joints. A continued pressure will break the fingers. This is a quick and effective method of escaping from this type of attack.

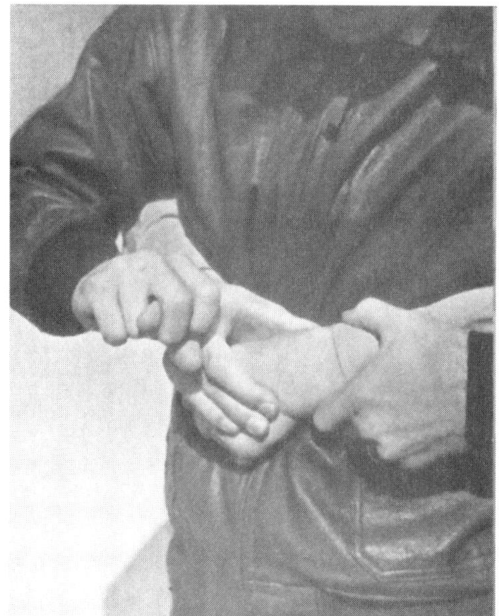

98. FOLDED FINGERS (REAR)

When body locked from the rear and the fingers are folded under so that it is difficult to grasp one of them, insert your fingers down between the knuckle of the little and ring fingers. Then pry outward. Note that the little finger is secured by your forefinger and thumb. This pressure will release his grasp. A continued outward pressure will break the finger.

I. Handlock

99. CHEST

The handlock is applied when an opponent places his hands on your chest. One hand secures the wrist from the underside, while the other hand is placed over the fingers, holding the hand firmly in place. A quick bend forward will force your adversary to his knees by chest pressure against his hand.

J. Wristlocks

100. REGULAR

The regular wristlock is applied with both hands. Your opponent's fingers are bent upward and the wrist is bent in. The twist is to the outside. The arm may now be pulled with an outward twist or pushed back with an outward twist. The arm is bent at the elbow.

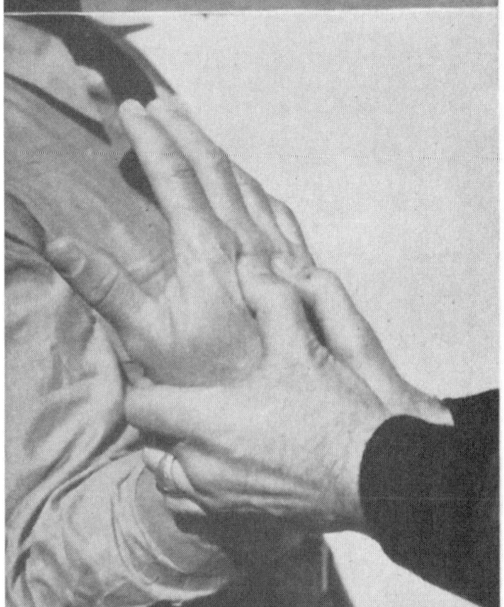

100A.

Note how the thumbs are placed on the back of the hand just below the third knuckles. The wrist is bent back at right angles to the forearm.

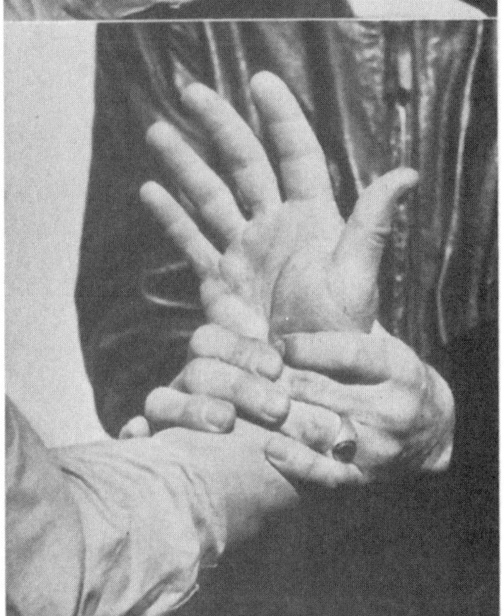

100B.

This illustration shows the inside of the wristlock. Note how the fingers are secured against the inside of the wrist.

100C.

This illustration shows how the wristlock may be secured when an opponent pushes you. Your left hand passes over the pushing hand and secures the thumb side. Opponent's hand, in this instance, is rolled with fingers up to the left.

101. REVERSE

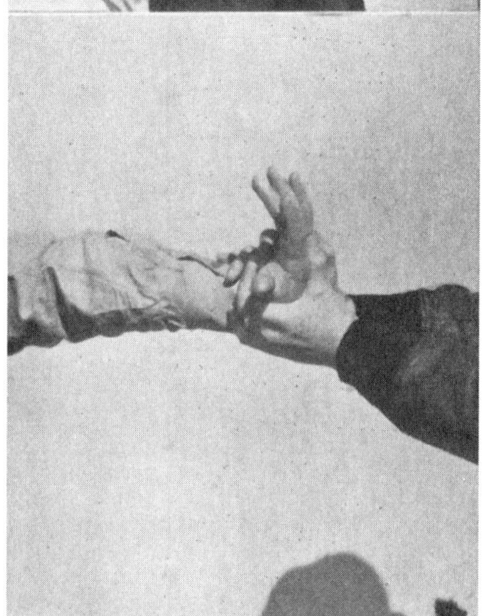

The reverse wristlock is applied and the hand and arm are turned just opposite to the regular wristlock. In this case the wrist and arm are rotated inward instead of outward. The arm is locked at the elbow in a straightened position. In the various forms of attack, it will be explained just how this wristlock is applied from many positions. Note that the thumbs are on the back of the hand and fingers are against the inside of wrist.

101A.

This illustration shows how you may kick the face of an opponent by stepping back from the position in illustration 101, using both hands to secure your opponent's wrist. Note the twist on the wrist and arm in this maneuver.

101B.

Pushing the arm upward with one hand and securing your opponent's belt with the other, you now have an effective "come-along."

101C.

Note that your hand passes over the pushing hand to the little finger side. Rotate the arm inward from here.

101 D.

This illustrates how the reverse wristlock may be applied by grasping an opponent's wrist while his arm is at his side.

102. DOUBLE

This wristlock is accomplished by clasping an opponent's wrist with the opposite hand (left hand clasps right wrist or right hand clasps left wrist). Your other hand and arm pass over his upper arm, under the bent elbow, securing your own wrist. This hold can be obtained from a variety of positions, but the manner of obtaining the hold always remains the same. The opponent's arm may now be jerked up the back into a twisting hammerlock if desired. (See illustration 112.)

102A.

This illustration shows the start of the double wristlock when an assailant is behind. Note that the right hand clasps the opponent's left wrist, and that your other arm is just beginning to pass over the opponent's arm above the elbow to secure your right wrist. Note, also, that your opponent's arm is pushed away from your body. With a quick jerk, his arm is forced outward and backward as you turn into him to the left.

102B.

Here, the start of the double wristlock is applied when an opponent attacks from a low position. Again, the wrist is secured by the opposite hand (right hand to left wrist). Your other hand is just beginning to pass over the secured arm above the elbow.

103. DOUBLE REVERSE

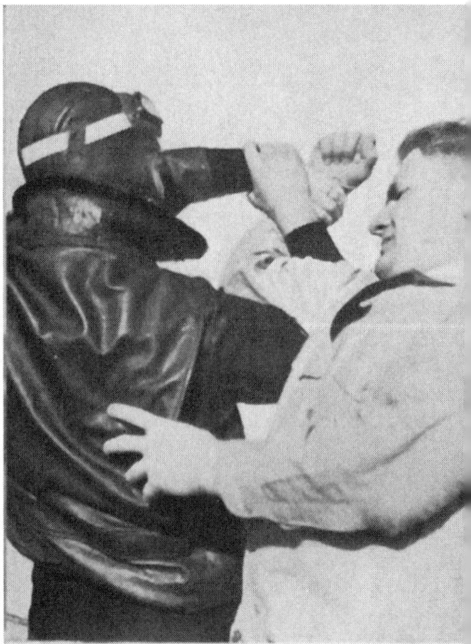

The double revere wristlock is obtained on an opponent's upraised arm. Here again, the left hand secures the right. The thumb side of the hand is down. Your other arm passes under the upper arm above the elbow and then over to secure your own wrist. Pressure which may snap the elbow or dislocate the shoulder can be exerted straight back, This is an effective maneuver against an assailant who attempts to strike out with a downward blow of a club, knife, or any other weapon.

K. Elbowlocks

104. SINGLE

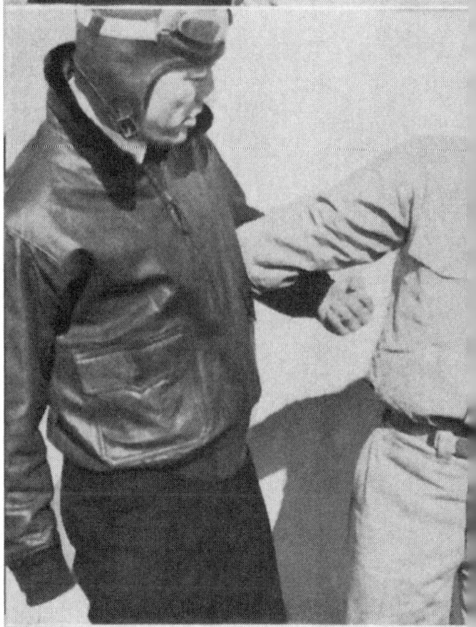

A single elbowlock secures the arm to the side effectively and allows freedom of the other hand and arm. The arm is locked in the bend of the elbow and against the ribs. Pressure is exerted against the elbow joint. An opponent must struggle strenuously to release himself. It will be noted that the hand of the arm which locks opponent's arm is not used, which would indicate the practicability of the maneuver, even though your hand were injured or otherwise useless.

105. DOUBLE

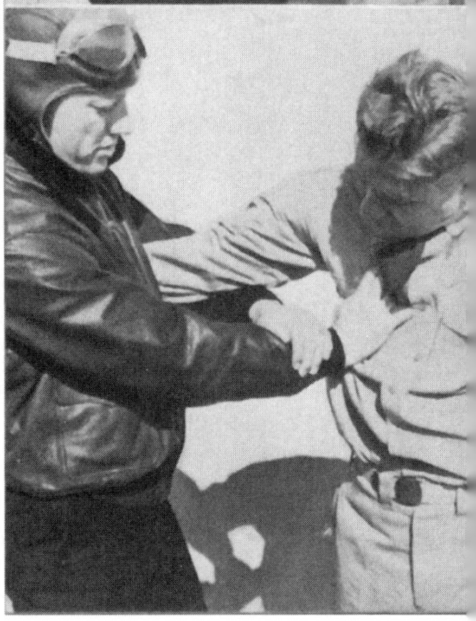

This elbowlock is obtained in the same manner as the single; however, both arms are used to lock your opponent more securely. In this instance, the locking arm gains more leverage by the hand clasping your other arm, the hand of which is placed against the opponent's chest. Again the leverage is against the elbow joint.

106. TWISTING

The twisting elbowlock is obtained when opponent's arm is bent at the elbow, the pressure being inward and upward against the elbow. This pressure can be applied with both hands and is done with a quick jerk of your arm.

107. OVER SHOULDER

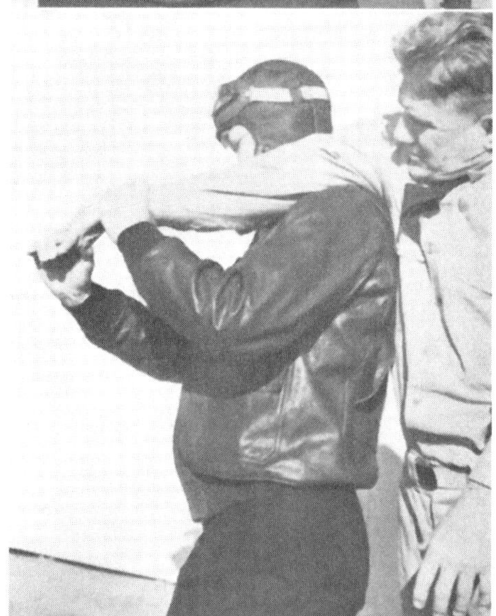

In this elbowlock, the elbow is fully extended over the shoulder. Pressure is obtained by turning the palm up, one hand securing the wrist and the other the fingers. A snap downward over the shoulder will break the elbow. The maneuver is effective over either shoulder.

L. Hammerlocks

108. REGULAR (WRIST UP)

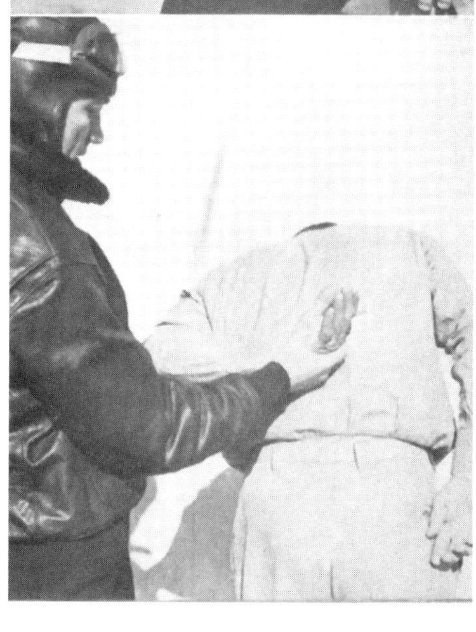

In this simple hammerlock, pressure is up on the wrist, while the elbow is secured to the side. To increase the pressure and pain, the forearm is forced up the back.

109. REGULAR (WRIST TWIST)

This hammerlock is exactly like the one above and is secured in a similar manner, with one exception. The wrist is twisted so the fingers point to the deck. Now, with one hand at the elbow and the other pushing against the back of the hand, a terrific pressure is set up. For additional leverage, force the forearm up the back. This tactic is very effective as a "lead" or for control of an unruly person.

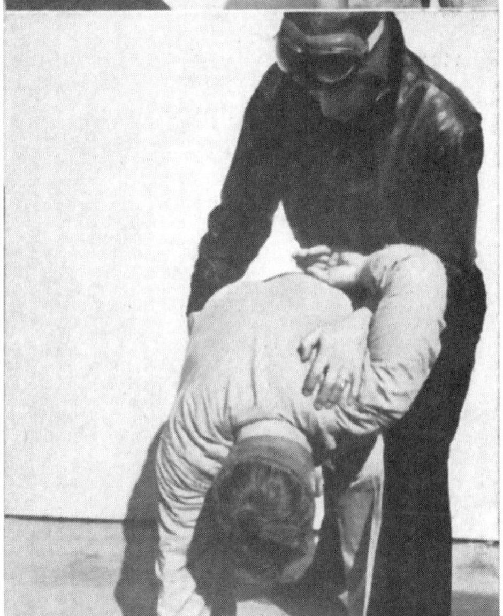

110. INSIDE BAR

Although this hold is somewhat difficult to establish, when once applied it is very effective. Note that your left arm secures your opponent's left arm; your hand is placed over the shoulder blade and passes under the forearm. Your body against the forearm prevents your opponent from throwing his arm to the left to release himself. When the inside bar hammerlock is applied, you are behind opponent, and pressure down on the shoulder and up against the forearm will force him to the deck in a helpless position. Your body must be in contact with his forearm at all times. Force him to deck quickly so that he does not get a chance to kick you.

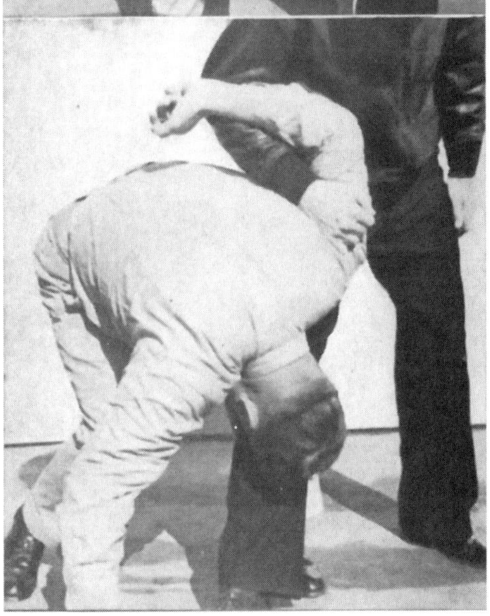

111. OUTSIDE BAR

When the outside bar hammerlock is applied, the right arm secures your opponent's left arm (or left secures right), the hand is placed at the shoulder, and pressure upward is applied against the opponent's forearm with the inside of your elbow. Hooking his forearm at your elbow prevents his escape; however, you must keep a continual pressure on his forearm to prevent this. A stiff pressure will force your opponent to the deck. Note that you are outside your foe's legs and are thus better protected from kicks. This pressure, as with the inside bar, will fracture the elbow or dislocate the shoulder.

112. TWISTING

The twisting hammerlock is the result of a double wristlock being pushed up the back and pulled out from the body. Note that hold on the wrist is the same as in illustration 102. This maneuver will break the arm at the elbow or dislocate the shoulder and is extremely dangerous. By a continued stiff pressure upward toward the head, an opponent can be driven to the deck.

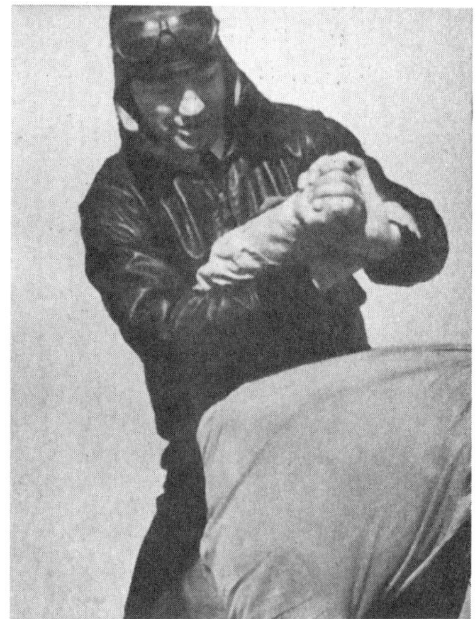

M. Shoulderlock

113. FROM DECK

When an opponent is brought to the deck on his face, forcing the arm upward and perpendicular to the deck, the wrist is twisted so the fingers point toward the head. This locks the shoulder and securely pinions the assailant. Added pressure will dislocate the shoulder. Both hands secure opponent's hand in this illustration, but one hand can do the job. Note that the elbow is not bent, but is extended and locked.

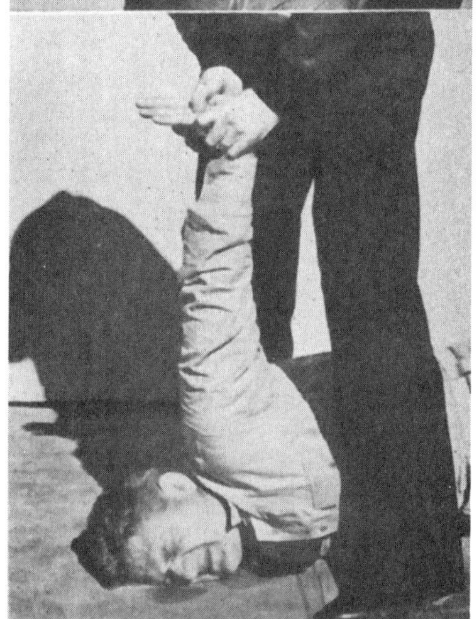

N. Necklocks

114. NECK SNAP

The neck snap is applied from close quarters, usually when an opponent has his arms about your body. Reach up behind opponent's head, secure his hair with your fingers, and snap the head sharply backward. This may cause serious injury to the neck, and always is very painful.

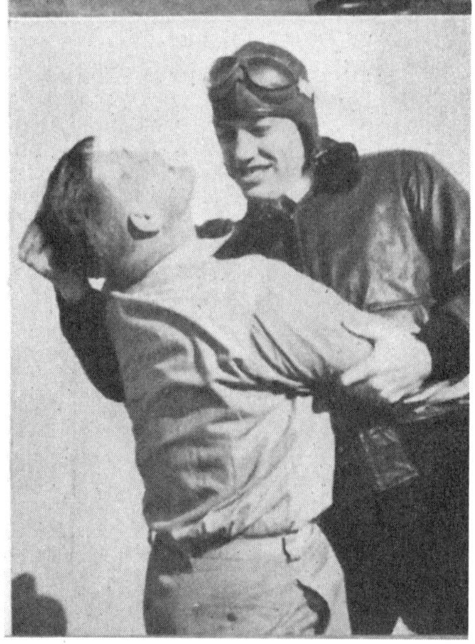

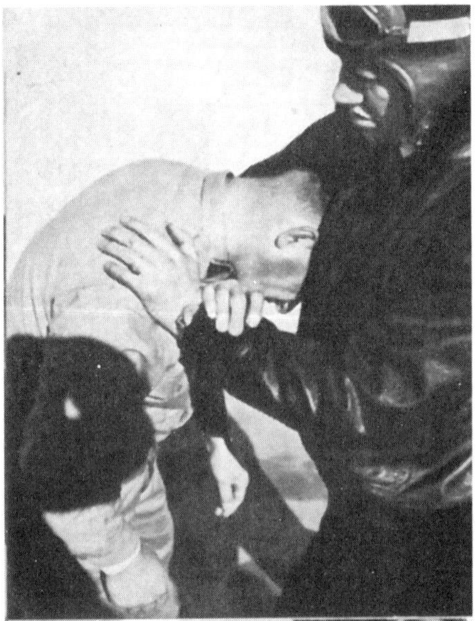

115. CHANCERY STRANGLE

This strangle hold is applied from the front and works effectively on any adversary who attacks with his head down. One arm is passed around the head and neck across the windpipe and locked over the other wrist or forearm. Pressure upward will strangle opponent. Note the hand on shoulder of the opponent. This hand could be placed in the armpit and thus be more secure, if desired. (See illustration 116.)

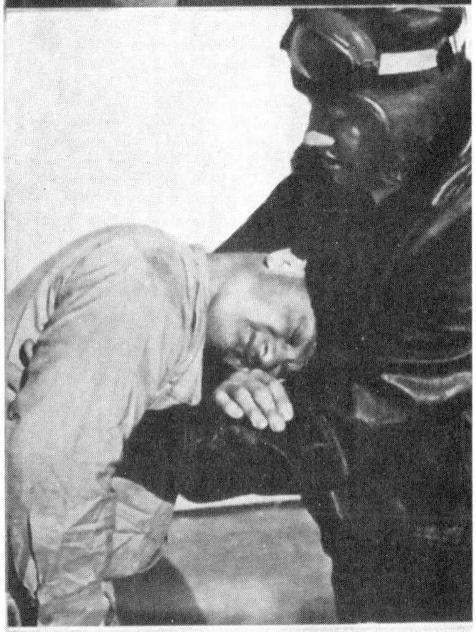

116. HEAD TWIST CHANCERY

This tactic is applied in the same manner as that in illustration 115, except that here the arm passes across the side of the neck. A quick snap upward will break the neck as all opponent's weight is put on the neck muscles. Note in this illustration that the arm and hand pass under the armpit. An opponent can easily be thrown with this maneuver by twisting quickly to the right.

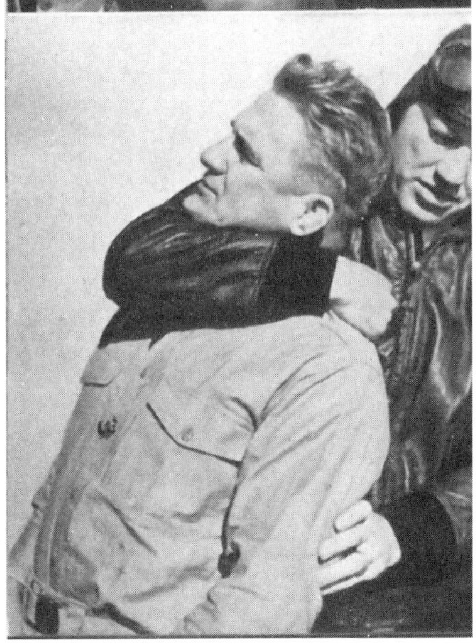

117. CLOSED STRANGLE

This illustrates the closed strangle. The neck is in the bend of the elbow. Your other arm is free for any defense or counteroffense that you wish to use. This tactic is desirable for use in only a few maneuvers and is not very effective, because it can be easily broken and leaves you open to various forms of attack. It is illustrated here, however, to acquaint you with a tactic commonly used.

118. LOCKED STRANGLE OVER THROAT

The locked strangle over the throat is applied by passing one arm over opponent's throat from the rear and locking the hand in the bend of the elbow of the other arm. The other arm and hand are placed on the back of opponent's neck, applying pressure forward, thus choking opponent. This strangle hold cannot be broken, and opponent will lose consciousness quickly

119. LOCKED STRANGLE—ACROSS SIDE OF NECK

The locked strangle across the side of the neck is applied as that in illustration 118. However, one arm passes across the side of the neck and not across the windpipe. Again the hand is secured at the elbow and the other passes behind the head. Pressure downward will cause your opponent to lose consciousness quickly. To keep from being struck in the groin or solar plexus, drop your opponent quickly to deck on top of you and lock your legs about him, so that he is held tightly against your body, thus he is unable to do any harm.

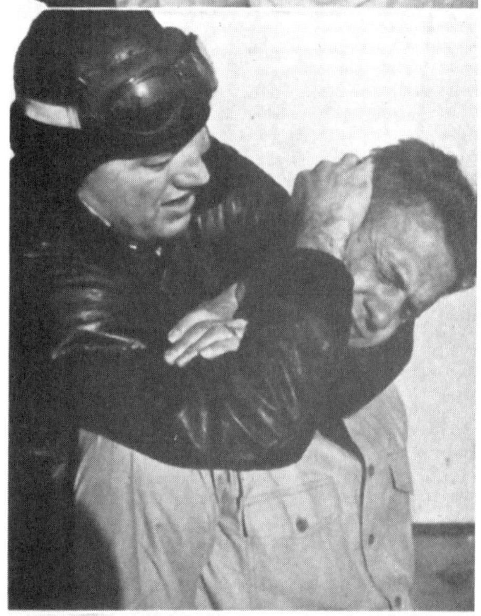

O. Hiplocks

120. REGULAR

The hiplocks are variations of Japanese jiu jitsu throws in which leverage and balance are used to overcome superior strength and weight. In this hiplock the opponent's arms are securely pinioned, the right with a single elbow lock and the left clasped at the elbow. Stepping across the front of your opponent's body, use the hip as a fulcrum to throw him to the deck on his back. His arms are still securely pinned and he, therefore, has no way of breaking the fall except with his feet. Note in the illustration how far the hip is pushed through; this is essential if the maneuver is to be worked properly.

121. REVERSE

Again the hip is used as a fulcrum to throw your opponent to the deck. In the reverse hiplock, opponent's arm is secured at the elbow with one hand and the other passes under the armpit and slightly across the back. Stepping directly behind your opponent and pushing the hip through allows you to pull your man over the hip easily. This maneuver causes your man to lose his balance and leverage.

122. REAR

This hiplock is applied when you are attacked from the rear. By twisting your body you pass your arm over opponent's head and bring it up under his arms and pinion his near arm above the elbow. By pushing the hip through and using it as a fulcrum, your opponent can be thrown heavily to the deck with your full weight across his chest. He cannot stop the fall.

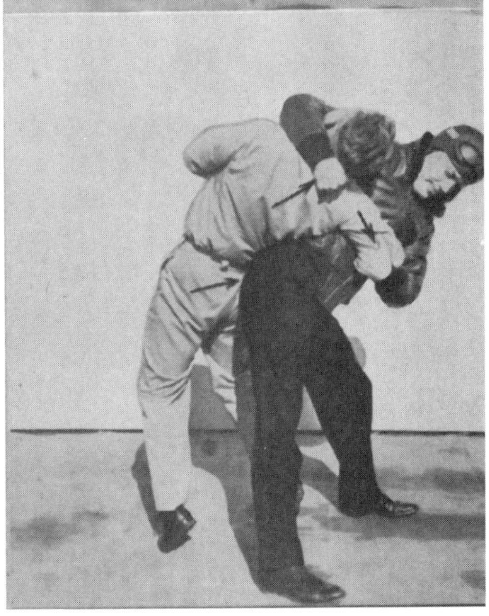

123. STRANGLE

This hiplock is worked the same as that in illustration 121 except that one arm passes across the throat instead of under the arm. Leverage is more effective here and allows you to throw your man with great force. Note again the hip position.

P. Leglocks

124. KNEE SNAP

Whenever an opponent's leg is straight, a blow at the knee is very severe. Here the foot is brought down on the knee with force. A shoulder or any other part of your body against the knee when it is in a straightened position will tear the ligaments and tendons causing very serious injury and completely incapacitating opponent.

125. STANDING TOE HOLD

This standing toe hold can be applied from many positions. The severity is produced by the twisting action against the toes which in turn puts a strain on the ankle. A quick twist will either sprain or break the ankle. Note how the leg is secured by one hand while the other applies the pressure.

126. BAR TOE HOLD

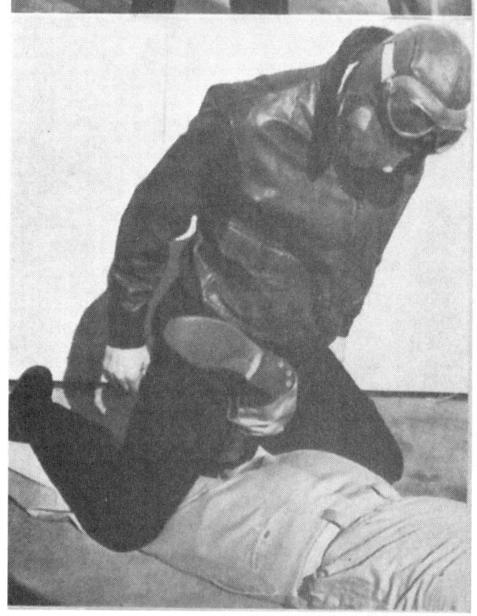

This maneuver is generally applied when an opponent is face down on the deck. Your leg is placed across the hock of your opponent's knee. His leg is then pulled up into your crotch. Thus the leg is locked and held securely by your body. Not only can you break the ankle or leg from this position with your hands, but you can attack any part of the back of your opponent. Note how you should sit astraddle your adversary. He is incapable of any movement if the pressure is applied with force.

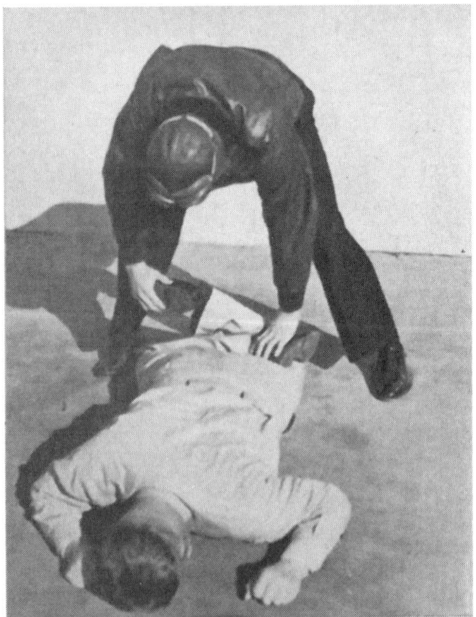

127. DOUBLE BAR TOE HOLD

The principle of leverage in this maneuver is the same as shown in illustration 126; however, you use our opponent's legs to gain the leverage. Note here how one leg at the ankle is placed behind the hock of the other knee.

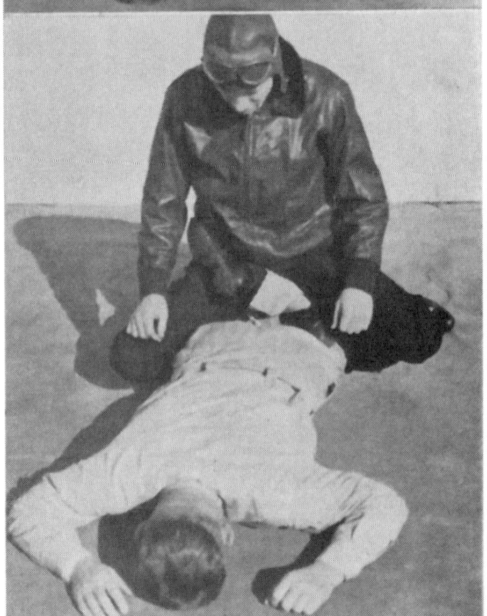

127A.
You now bring his ankle up and sit astraddle the lever leg. The greater your body pressure forward, the less your opponent is able to move.

Q. Flying Mare

128. USE HIPS
In this flying mare, first firmly grasp opponent's arm or arms, spread the feet wide apart and throw him over your head by a quick bend at the hips. The legs are spread to afford the greatest amount of leverage.

129. USE KNEE

In this flying mare the arm is secured firmly. Dropping to one knee and bending quickly forward, you easily throw your opponent over your head to the deck. This maneuver is the better of the two flying mares herein illustrated. Your leverage and balance are more secure.

CHAPTER 6

Frontal Attack

THE TACTICS discussed here are listed and described in the manner the attack is likely to occur. First, rushing; second, pushing and striking; third, grasping; fourth, strangling; and fifth, body locking.

It will be noted that these maneuvers are for use against an unarmed opponent. Defense and counterattack against an armed adversary will be discussed in a later chapter. At the outset, it should be emphasized that it is not necessary to wait for an enemy to attack first. Each tactic in itself can become an effective offensive maneuver.

It also will be noted in practically every maneuver that you close in on your enemy in much the same way as a boxer clinches. This cuts down his effectiveness.

The instructor need not necessarily follow the order of the maneuvers as they are set up here; he may make adjustments to suit his teaching program. However, it is urged that maneuvers themselves be taught as they are illustrated. Slight variations due to a man's size and weight are permissible as long as they do not destroy the effectiveness of the tactic.

In all frontal attack maneuvers it is necessary to be on the alert for any type of attack devised by your foe. Alertness is just as important in Hand-to-Hand Combat as it is to the boxer or wrestler in the ring. Therefore, it is reemphasized that stance is most important. The hands, feet and body should always be in such a position as to react in the shortest time possible. In other words, you cannot stand flat-footed with arms at sides but must assume a position of readiness—a slightly crouched position, legs slightly bent and hands active where they can be brought into play immediately for protection of the head or body.

A. Rushing

The maneuvers and tactics listed below, and under succeeding sections, should be studied in the progression they occur. In order to prevent obvious repetition, however, it will be noted that a reference is made only to maneuvers and tactics previously illustrated, but which, nevertheless, should be thoroughly understood. The new maneuvers and tactics described and illustrated in the following pages should be studied with this thought in mind. Thus, the student will have a complete understanding of all the methods available to him in any form of attack through a careful review, plus the accomplishment of new techniques.

1. KICKS (*See illustrations 62–68*)

As an opponent rushes, any of these kicks can be used against him effectively. Deliver your kicks to the area vulnerable to attack at the moment. This depends on, of course, the position of your opponent as he rushes you. The kick illustrated in 68 may be used as you side step him.

2. HAND BLOWS (*See illustrations 4–15, 18–27, 29, 30, 31*)

Again the desirability of using any particular maneuver depends largely on the manner your opponent rushes you. Therefore, you should become familiar with all the hand blows listed above and use them with force when the occasion demands it. The hand blows illustrated in 19 and 31 can be used to your opponent's rear as you side step him.

3. HEAD TWIST CHANCERY (*Note illustration 116, and picture series 130*)

4. CHANCERY STRANGLE (*Note illustration 115*)

This maneuver is executed exactly like the head twist chancery above, except that your arm passes directly across the throat causing strangulation. Kicks, kneelifts, strangle, snap neck or throw to deck can be executed.

5. ARM DRAG (*See picture series 131*)

6. SWITCH (*See picture series 132*)

Head Twist Chancery

130A—Opponent rushes from a crouched position to attack below the waist.
130B—Using one arm to ward opponent off, pass the other arm about his neck, twisting his head sideways.
130C—Place the arm used to ward him off in his armpit, through to shoulder blade, and lock the arm which passes across his neck in the bend of the elbow. This locks your opponent's head inside your arm. Raise up with the arm under his armpit and twist down on the neck.
130D—You are now in position to bring our foot into his groin, kneelift to the stomach, or jerk quickly upward placing all his weight on his neck, thus snapping it easily. Opponent can also be thrown to deck by continuing your twist and you will fall across his chest.

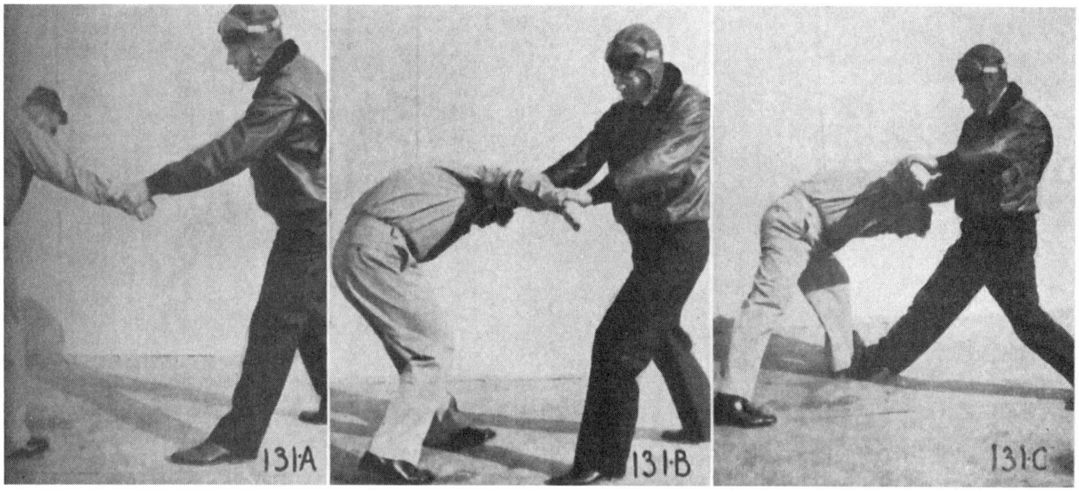

Arm Drag

131A—As your opponent rushes, reach straight across and grasp his arm at the wrist.
131B—Your other arm clasps his upper arm from the underside, thus keeping his arm straight.
131C—Throw your leg in front of your opponent, step on his opposite arch and begin your drag backward. This trips your man.
131D—Fall backward pulling opponent to side so that he falls to deck and not on you. A quick jerk on the arm drives your opponent's head onto the deck. He is unable to prevent himself from falling, due to your tripping leg and pressure on the arm, which should be locked straight.
131E—This illustration indicates how the same maneuver may be executed by grasping the clothing of the upper arm or shoulder.

Switch

132A—Again the arm is secured by reaching straight across while your other hand is inserted in your opponent's crotch at the same instant your inside leg hooks inside opponent's foot.

132B—By rearing back and using the arm and hand inserted at the crotch as a leverage, force opponent to deck.

132C—Continuing this pressure forces opponent head first to deck. Turning inward brings you on top of your victim and in position to continue the counterattack with the hands.

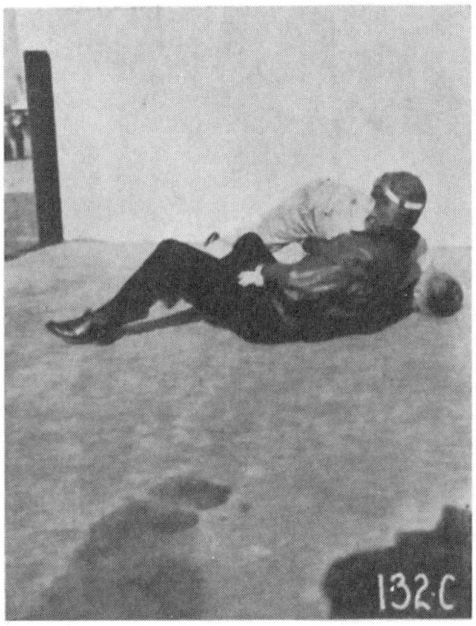

B. Pushing and Striking

(Maneuvers are listed in the order they should be studied. Refer back or forward as the case may he so a thorough understanding of the maneuvers in the proper sequence is obtainable.)

1. KICKS (*See illustrations 62–67, 73 and 124*)

An opponent is always open for these listed kicks when he is within range of a push or strike. Deliver them with force, depending on the position of your adversary.

2. HAND BLOWS (*See illustrations 4–15, 18, 20–27, 29–33, and 35*)

The approach or position of opponent will determine which type of maneuver to use. Learn them all and be ready at all times to use the one most practical.

3. ELBOW BLOWS (*See illustrations 40–44*)

Always make good use of your elbow whenever a situation at close or comparatively close quarters warrants it. The elbow is a vicious weapon; therefore, you should become familiar with its use.

4. KNEELIFTS (*See illustrations 50–54*)

Bringing the knee up into a vital or vulnerable area, especially the groin, will generally bring about the conclusion of any attack. Again the area struck depends on your opponent's position during attack.

5. HANDLOCK (*Note illustration 99, and picture series 133*)

6. FINGERLOCKS (*Note illustrations 93–95, and picture series 134*)

7. REGULAR WRISTLOCK (Note illustrations 100, 100A, 100B and 100C)

a. Outward Twist (*See picture series 135*)
b. Pull Inward, Backward Twist (*See picture series 136*)

8. REVERSE WRISTLOCK (*Note illustrations 101, 101A, 101B, 101C, 101D, and picture series 137*)

9. ELBOWLOCKS

a. Single Elbowlock with trip (*See illustration 104*)

Locking the elbow with a single arm, put foot behind opponent and trip him back over it. This can be done by turning toward the locked elbow.

b. Double Elbowlock with trip (*See illustration 105*)

The elbow is double-locked and the trip is executed in the same manner as above.

c. Twisting Elbowlock (*See illustration 106*)

Obtain your elbow lock on the arm with a quick jerk. Then turn quickly and throw your leg behind, tripping your opponent over it.

10. HEAD TWIST CHANCERY (*See picture series 130*)

This maneuver is executed exactly as that illustrated under 130. Naturally you are at closer quarters; however, this will not hamper, but enhance your ability to work the tactic.

11. CHANCERY STRANGLE (*See picture series 130 and illustration 115*)

This tactic is executed as head twist chancery, except that the arm passes over the throat to strangle.

12. ARM DRAG (*See picture series 131*)

Maneuver is executed in same manner as illustrated in rushing attack.

13. SWITCH (*See picture series 132*)

This tactic is accomplished in same manner as when rushed by opponent.

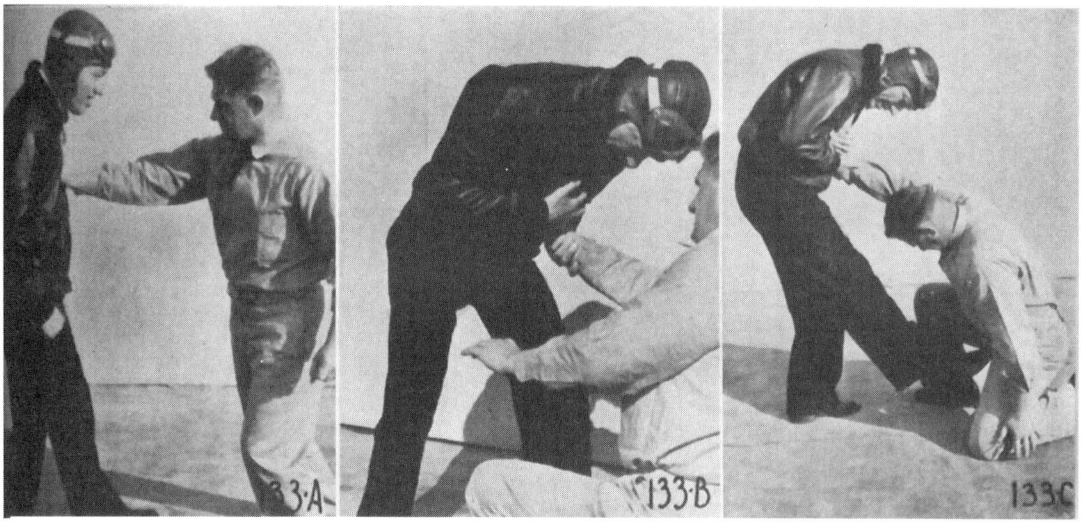

Handlock
133A—Opponent pushes with the hand open against your chest.
133B—One of your hands comes up and secures your opponent's wrist from the underside, while the other hand presses his finger firmly against your chest. By bending forward at the hips, you bring your opponent to his knees.
133C—This allows toe kicks to the groin or stomach. If you now desire to release your grasp, step back and kick to the chin or face.

Fingerlocks
134A—As your opponent strikes or pushes you, secure his fingers in one of the fingerlocks listed in illustrations 93 or 94. Note how the other hand secures the wrist. If you wish, secure the fingers as in illustration 95 with a finger spread and begin your pressure up and back.
134B—This brings your man to his knees where kicks can be used to the groin or other vital areas.

Regular Wristlock—Outward Twist

135A—Opponent begins his push or strike. Note how your hand should pass over the outstretched arm and secure opponent's hand on the thumb side.
135B—Opponent's hand is secured at the wrist proper with both hands. The thumbs are against the back of the hand and the fingers pass over the inside of the wrist. (See illustrations 100A and 100B.) You now begin your outward twist and step back at the same instant.
135C—This forces opponent to the deck on his back or side depending on your hand pressure. From this position, you can drive his elbow to the deck, breaking it; press straight down on the wrist, fracturing it; or—
135D—Jerking opponent up to his side, attack his head, neck, spine, kidneys with all types of kicks as in illustrations 71, 72, 80, 81, or 82.

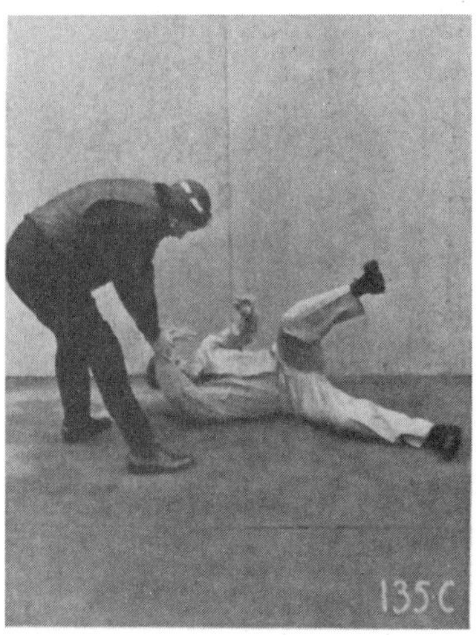

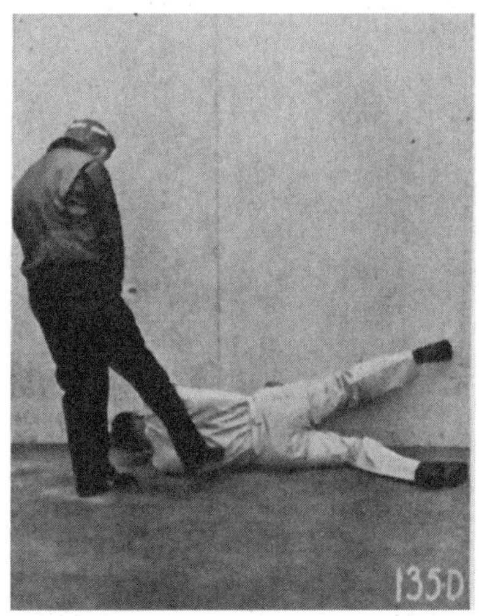

Regular Wristlock-Pull Inward, Backward Twist

136A—Secure hand as in illustration 135A.

136B—Pull forward with both hands securing wrist proper (see illustrations 100A and 100B) to the inside. This causes opponent to come forward.

136C—As opponent steps forward or leans forward, reverse your action and twist overhead backward with a quick jerk.

136D—The pressure created on the arm is great and may cause it to be broken due to the forward momentum of your opponent. You may force your foe to the deck and attack with kicks or use knee lifts to the groin.

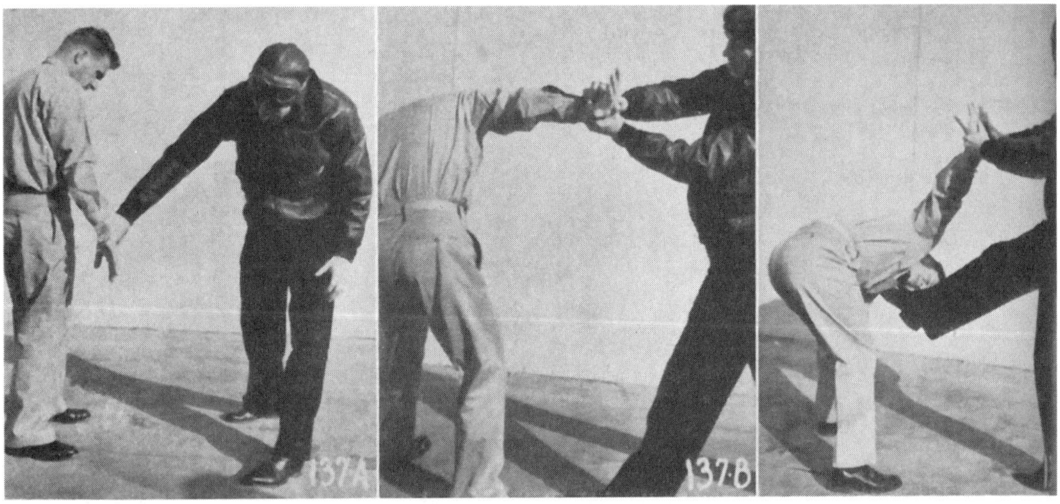

Reverse Wristlock

137A—It will be noted that the hand is secured on the little finger side as in illustration 101C. This is true if your assailant struck or pushed at you.

137B—Now, turning the wrist in the reverse direction of the regular wristlock and with both of your hands in the proper position (thumbs to back of hand and fingers across inside of wrist), jerk the arm upward.

137C—From this position you can deliver kicks to the face, chest, or solar plexus. You can bring your opponent's face within range by pressure of the thumbs against the back of the hand.

137D—By placing one hand on the elbow to keep it locked, you can control your opponent and keep him from turning.

137E—Still securing the upraised arm with the fingers turned toward head from position in illustration 137D, step between opponent's arm and body. Putting pressure on elbow with other hand, you can force your opponent to the deck and attack with kicks or knee drops.

137F—If desired, you can push the wrist high and grasp the belt for an effective "come-along."

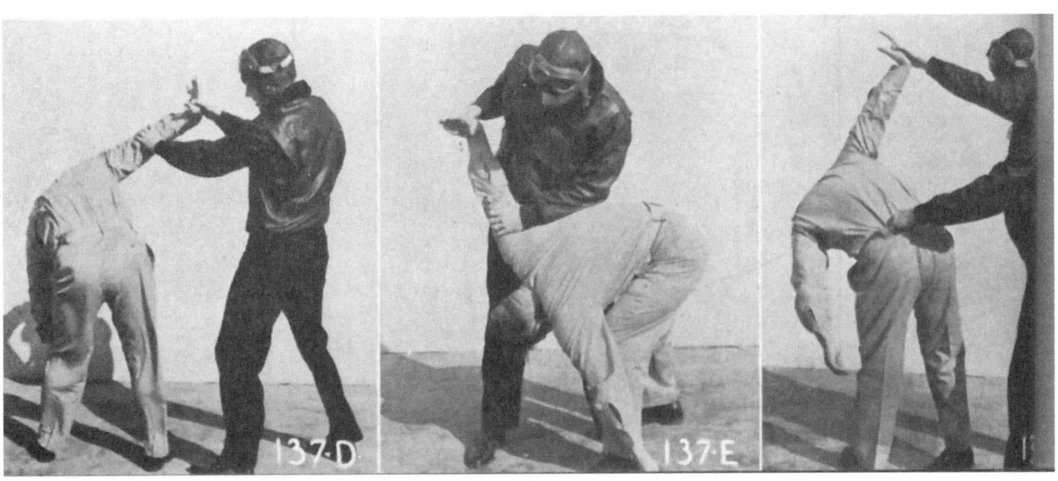

C. Grasping

(Study maneuvers in order listed, referring back or forward as the case may be.)

1. KICKS (*See illustrations 62–67, 73 and 124*)

Kick effectively to those areas open to attack, depending on the line of attack of adversary. Know and practice these kicking blows; get in the habit of using your feet.

2. HAND BLOWS (*See illustrations 4–15, 18–33, 35*)

Develop a proficiency in the use of your hands. Study the illustrations and practice the techniques. Don't be afraid to strike out when you are in a tough spot, but know where to strike and what to strike so you don't injure yourself.

3. ELBOW BLOWS (*See illustrations 40–44*)

Form the habit of using your elbows. You can use them on occasions when a hand blow is ineffective. Snap them hard!

4. KNEELIFTS (*See illustrations 50–54*)

Get the habit of lifting the knee when attacked. It is a very effective and efficient weapon. Study the photographs and you will see what we mean.

5. RELEASES

a. Against Single Hand Grasp (*See picture series 138*)
b. Against Double Hand Grasp (*See picture series 139*)

6. REGULAR WRISTLOCKS (*See picture series 135, 136 and 140*)

The two regular wristlock combinations will work just as well from a grasping attack as from a pushing or striking attack. It is necessary, however, to pull off the fingers from the arm, clothing, etc. Study the picture series thoroughly.

7. REVERSE WRISTLOCKS (*See picture series 137 and 141*)

a. The reverse wristlock is applicable here as well as from a striking or pushing attack with one exception; the hand must be peeled from the grasp first.
b. Grasping Hair (*See picture series 142*)

8. ROLL AGAINST ELBOW (Outboard)

This tactic is very simple. Roll outboard against the elbow joint, thus locking it and easily escaping any grasping hold. Your adversary must let go or receive a fractured or severely sprained arm.

9. DOUBLE WRISTLOCK (*See picture series 143*)

10. CROTCH PICKUP

This maneuver is practicable when your opponent is in position 143A. Pass your near arm in the crotch from the front and far arm into the crotch from the rear. Secure hands together, lift opponent from deck, and bounce on head. As opponent is raised from deck, he loses his balance. Forcing your legs backward as you secure his crotch will break his grip on your legs.

11. FINGERLOCKS (*See illustrations 93–95 and picture series 134 and 144*)

Fingerlocks are used in grasping maneuvers as well as in striking or pushing. However, it may be necessary to peel finger loose on occasion. Review the past series and study techniques in picture series 144.

12. REGULAR HIPLOCK (*See illustration 120 and picture series 145*)

NOTE: It would be well to learn to throw your feet over quickly so that you can break your fall if you are ever caught in a hiplock throw. Throwing the body with the maneuver rather than resisting it, betters your own chances of escaping serious injury. Practice going with the throw and try to catch yourself with your feet.

13. REVERSE HIPLOCK (*See picture series 146*)

14. BACK FLIP (*See picture series 147*)

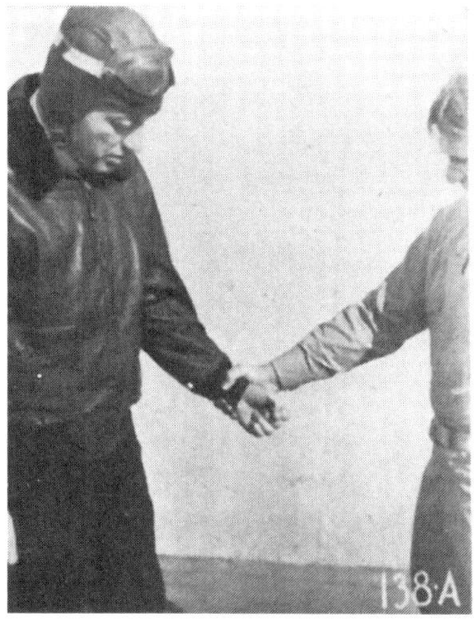

Releases—Against Single Hand Grasp

138A—When you are secured as illustrated by the clothing or lower arm, turn your opponent's hand, palm up—

138B—And roll your arm or wrist against the thumb side. As the thumb is the weakest point in the grip, the hold is easily broken.

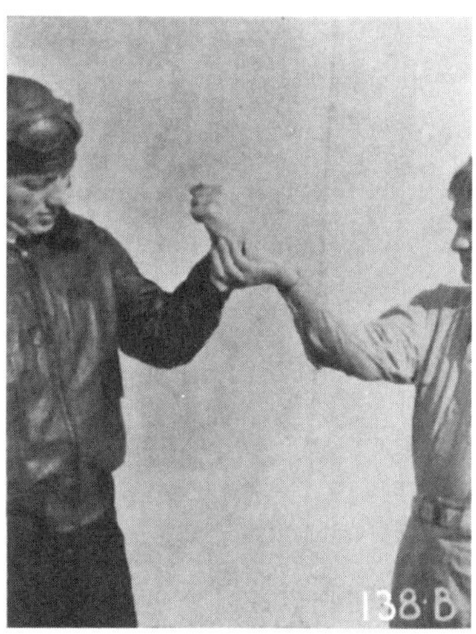

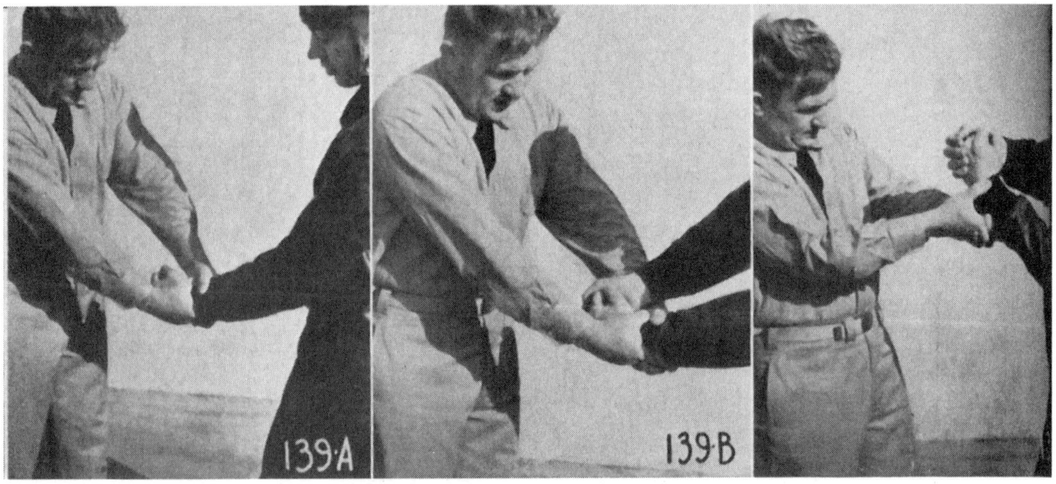

Releases—Against Double Hand Grasp

139A—Note the double grasp on the wrist by opponent.
139B—Reach over and down between opponent's wrist and grasp own fist firmly. Use opponent's wrist as a lever and begin your upward pull.
139C—A continued upward pressure against your opponent's thumbs releases you from this grasp with little or no effort.

Regular Wristlocks

140A—Opponent has grasped the hair.
140B—Reach over the grasping hand to the thumb side, secure the hand, and peel it off. Now follow out the regular wristlock procedure as described previously. (Picture series 135 or 136)

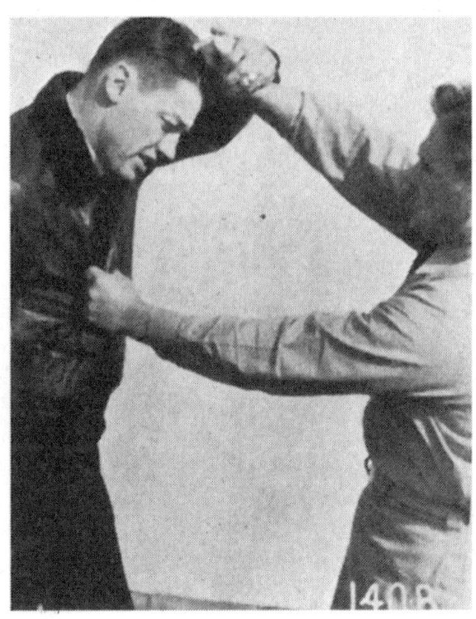

Reverse Wristlocks

141A—Opponent has seized the arm at the elbow.

141B—With your other hand, grasp the little finger side of the hand and peel it off. If the grasp is very strong, apply additional pressure to the little finger to pry loose.

141C—You now go into this hold by stepping between the arm and body, exerting a downward pressure against the shoulder and forcing your opponent to the deck. Keep the elbow rigid by turning the wrist so that the fingers point toward victim's head. Use kicks to face to subdue if desired. Follow the regular procedure in obtaining the maneuver, but do it quickly; don't waste time.

Reverse Wristlocks—Grasping Hair

142A—Opponent grasps hair.

142B—Reach over and clasp the little finger side of opponent's hand, and for additional leverage secure his elbow. Peel his hand off.

142C—Turning into his grasping arm, twist the wrist, keeping the elbow rigid. You can now go into the position in illustration 141C. Note how your leverage throws opponent out of position and off balance. In addition, he is in no position to attack further. Naturally, all these types of maneuvers must be worked rapidly to be effective. Of all maneuvers, the reverse wristlock is probably the easiest to accomplish because you work against the weak muscles in the arm.

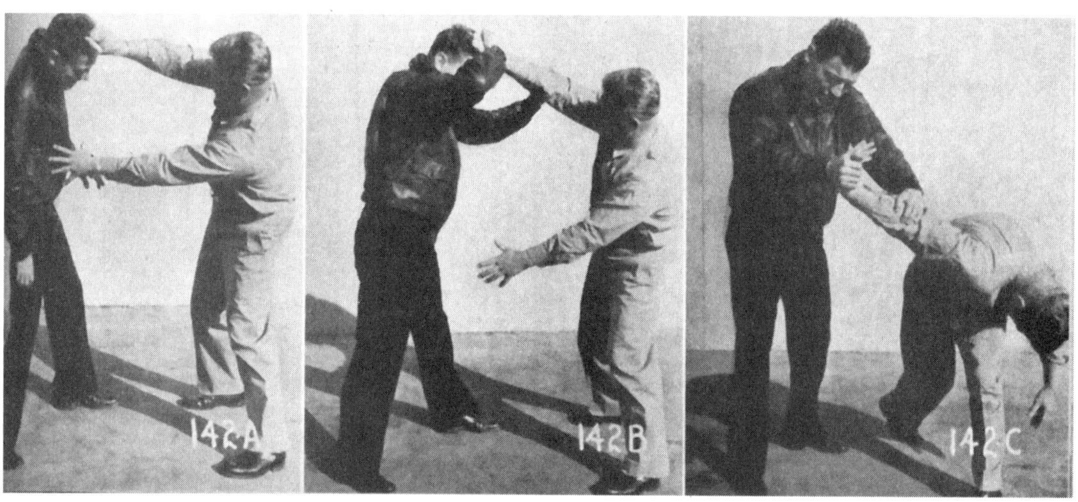

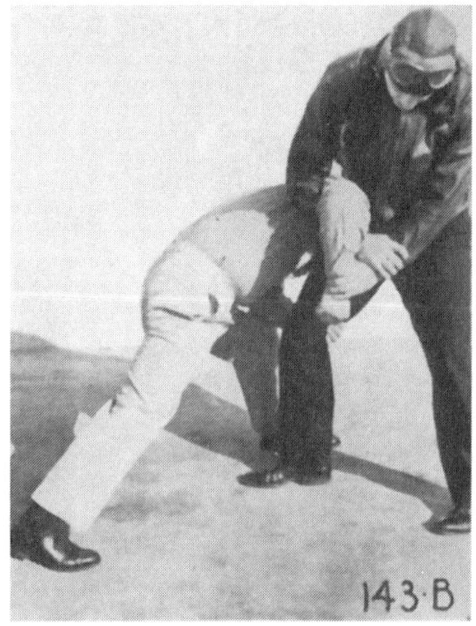

Double Wristlock

143A—As opponent grasps you about the legs, reach straight down and seize his wrist, fingers across back of hand, and thumb across inside of wrist.

143B—Snapping opponent's arm loose, pass your other arm over his upper arm and then through the bend at the elbow, securing it to your own wrist.

143C—Step back with your far leg and force this locked arm up his back and away from his body.

143D—Continued pressure up the back will force opponent to the deck. Additional pressure on the wrist will fracture the elbow or dislocate the shoulder. Notice how the pressure forces opponent's head against deck.

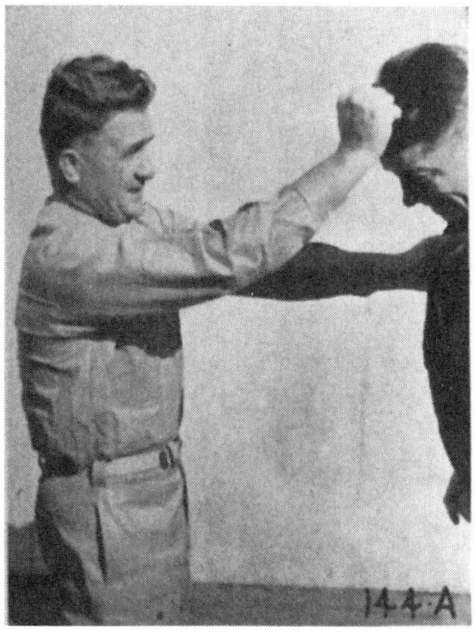
144·A

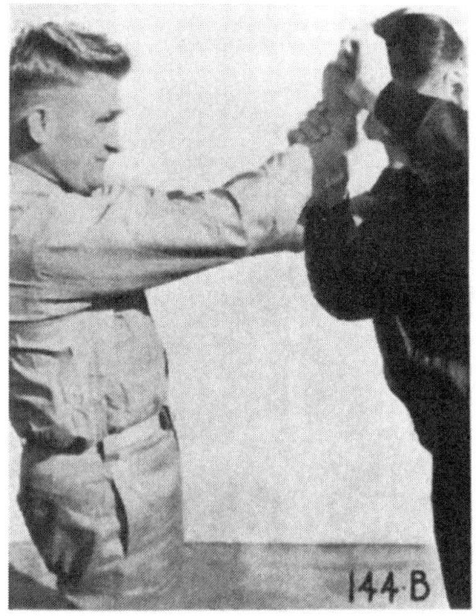
144·B

Fingerlocks

144A—Opponent has grasped hair.

144B—With one hand secure wrist from underside and peel off a finger with other.

144C—Applying pressure out and down while maintaining your grip on the wrist, force opponent to his knees.

144D—Carry on with a knee lift to the chin or face, or kick to the groin or solar plexus. By twisting outward you can bring your opponent to deck for a variety of kicks to vulnerable areas.

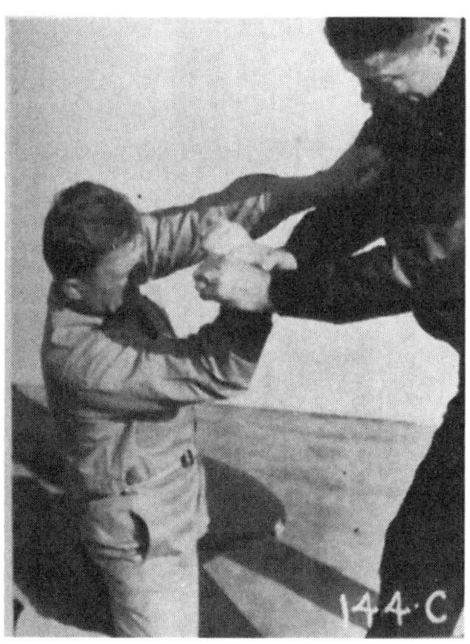
144·C

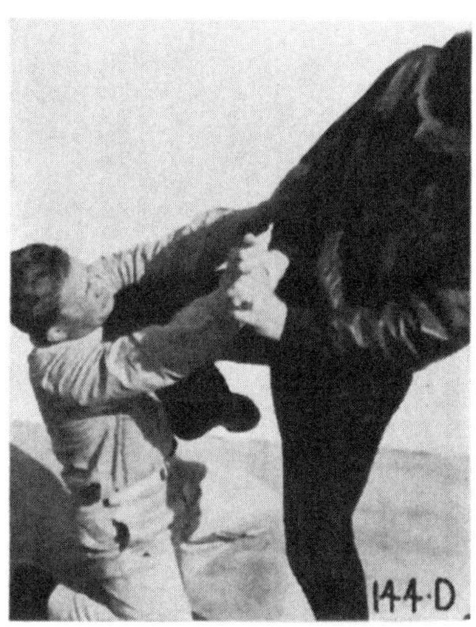
144·D

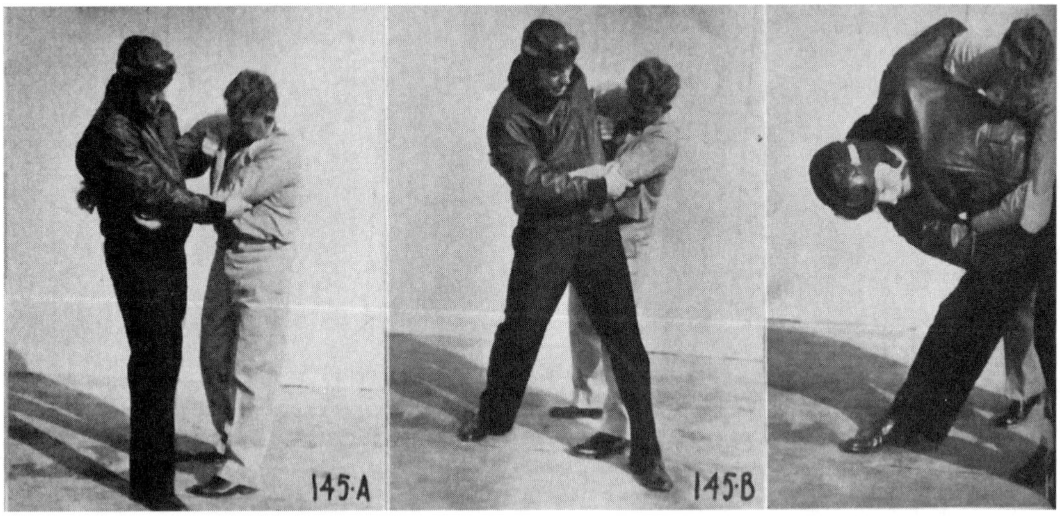

Regular Hiplock

145A—From a grasping position you have secured your opponent. Your far arm elbow locks his far arm and is held securely in bend of elbow, against your side. Your near hand secures his near elbow. Thus, your man is now held in an upright position.
145B—Keeping your arm hold, bring your far leg across your opponent's body.
145C—Still maintaining your hold on his arms, bend forward, and using your hip as a fulcrum, raise opponent from the deck.
145D—With a quick pull downward, throw your opponent over the hip in an arc.
145E—Your opponent falls to his back on the deck with your full weight upon him; or you may release your grasp slightly and use a knee drop to the chest or solar plexus, causing serious injury. (See illustrations 58, 59 and 60)

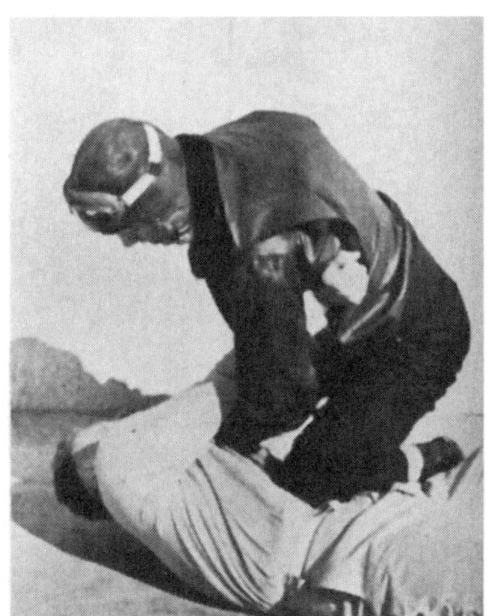

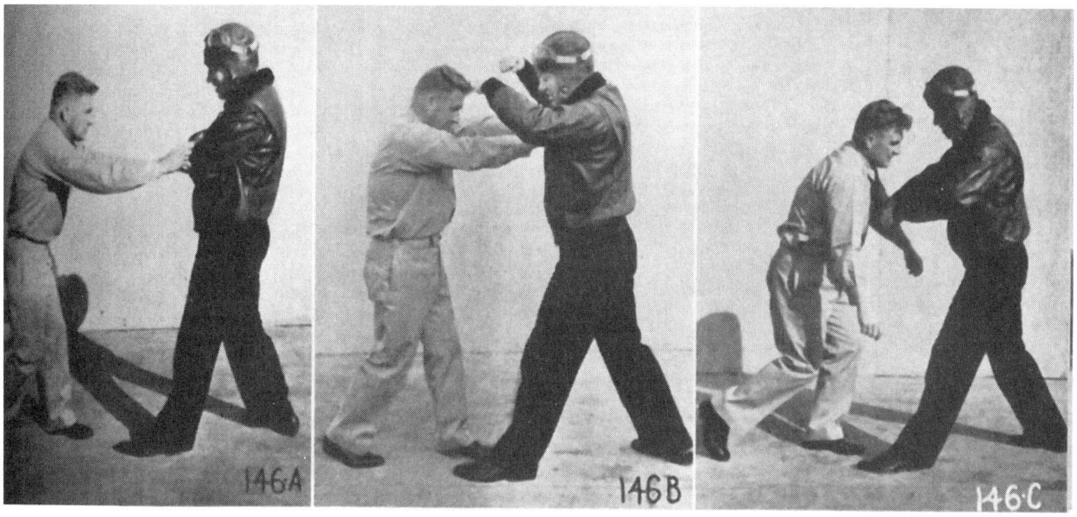

Reverse Hiplock

146A—Opponent is attempting to grasp or does grasp.
146B—Knock his arms loose with downward blows of both arms.
146C—This will throw opponent's arm downward to this position.
146D—Quickly place the far arm under opponent's armpit with the hand at the shoulder blade and secure the near arm just above the elbow with the near hand.
146E—Maintaining this hold, place your leg and hip behind opponent's and begin pulling him backward over the hip. The hip acts as a fulcrum and aids in lifting opponent from deck.
146F—Drop opponent on his head or back of neck causing injury to either. Again, use caution and when practicing this tactic, learn to throw your body with the maneuver, breaking your fall with your feet. Attack may be continued with knee drops or kicks.

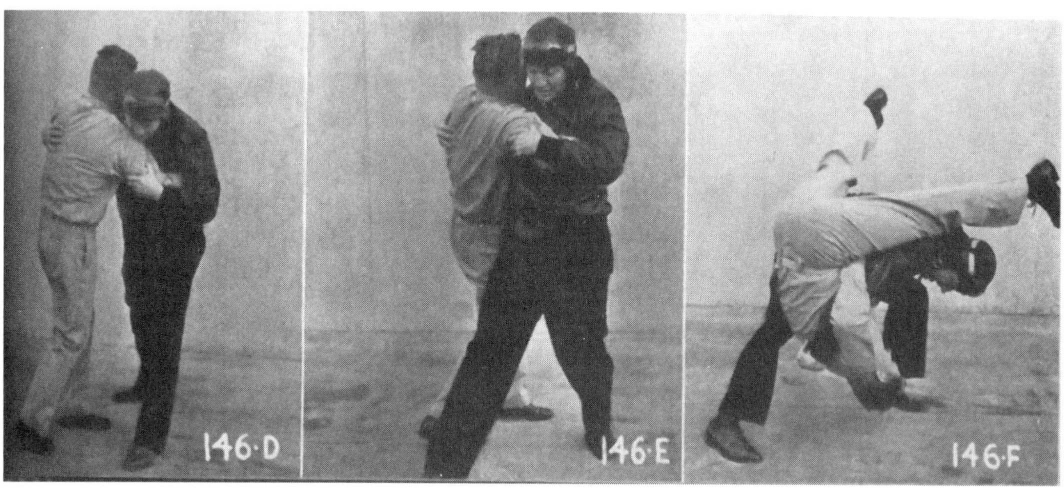

Back Flip

147A—When opponent grasps, you grasp him by the lapels or by clothing at the shoulders. Now jerk him toward you.
147B—As he comes toward you, place your foot in his stomach or chest, still retaining your grip on his clothing.
147C—Fall backward retaining clothing hold and keeping foot in place.
147D—Continue the backward movement and use your leg to throw opponent over your head. From this position you may release your grasp on the clothing and pitch him over with your leg, or retain your grip as you pitch him with the leg and use it for a means of control after he hits the deck. This maneuver works best when your opponent is coming toward you. His momentum helps carry him over.

D. Strangling

(Study maneuvers as listed referring to illustrations as needed.)

I. Arms Length

1. KICKS (*See illustrations 62–65, 73*)
If an opponent is strangling at arm's length from the front, he is susceptible to the many kicking blows illustrated in this manual. This type of attack is dangerous to an opponent.

2. HAND BLOWS (*See illustrations 4–15, 20–28, 32, 33*)
Many types of hand blows are available to you when attacked with an arm's length strangle. Be sure you are familiar with all of them.

3. ELBOW BLOWS (*See illustrations 40–44*)
Devastating blows can be delivered from this position. Learn them all thoroughly.

4. GRASPING HAND (*See illustrations 36–39*)
When in position, use your hands to grasp the anatomy as illustrated, also the nose or ears. Dig in with your fingernails.

5. NERVE SHOCKS (*See Illustrations 35, 83, 85–91 and charts in Chapter 4*)
You cannot kill a man with nerve shocks, and often in the heat of conflict they are difficult to find. However, it is well to know how to use them, as they will cause an opponent to flinch or release his grip or grasp. Some are more severe than others. Learn them; try them on yourself and see how they affect you. Pay particular attention to the nerve shock listed in illustration 90.

6. FINGERLOCKS (*See illustrations 93–95, and picture series 134*)
Peel off the fingers and gain your fingerlocks as learned previously. The full maneuvers are worked exactly the same.

7. REGULAR WRISTLOCK (*See picture series 135 and 136*)
From this form of attack, the regular wristlocks work with the same ease. It may become necessary to peel the hand from its grasp, but nevertheless the hand action remains the same.

8. REVERSE WRISTLOCK (*See picture series 137 and 148*)
Use the reverse wristlocks from this position in the same manner as from other types of frontal attack. Also study the maneuver below.

9. ELBOWLOCK (*See picture series 149*)

10. REGULAR HIPLOCK (*See illustration 120 and picture series 145*)

This hiplock can be started from a strangling position by bringing both arms up between your opponent's arms. You then secure them as shown in illustration 145A and carry out the rest of the maneuver. In order to break the grasp on your neck, bend your knees and bring your arms up between your opponent's with force.

11. REVERSE HIPLOCK (*See picture series 146*)

This reverse hiplock can be worked easily if you first break the hold on your throat by bringing your arms up quickly between your opponent's and obtaining the position shown in illustration 146D. You are now in position to carry on with the rest of the maneuver.

12. STRANGLE HIPLOCK (*See picture series 150*)

13. ARM WEDGE AND OTHER ESCAPES (*Note illustrations 4–11, 14, 15, 21–26, 28, 32, 33, and picture series 151 and 152*)

It will be noted that there are a variety of maneuvers listed here. Only a few of the blows are illustrated, but the student can readily see what other blows could be used by studying these illustrations. It can be readily seen that the arm wedge is a valuable maneuver to know since this same tactic can be used to overcome an adversary who has you pinned to the deck and is trying to strangle you to death.

14. OUTSIDE BLOW TO LOCKED ELBOW (*See picture series 153*)

Notice in these tactics that your opponent is always open for groin kicks. Don't hesitate to use them. Whenever strangled at arms length, the first thing you always must do is to tighten your neck muscles to resist pressure as much as possible. Do not work the tactic slowly or deliberately, but use speed and strength

II. Closed

1. KICKS (*See illustrations 62, 63, 73*)

Due to the close quarters, kicking maneuvers are limited. However, those referred to can be used effectively.

2. HAND BLOWS (*See illustrations 13, 16, 19, 28, 32–34*)

Hand blows are limited because of the proximity of the combatants; however, the blows referred to above are excellent for use when one is locked in a closed strangle.

3. GRASPING HAND (*See illustrations 36–39*)

It has been mentioned before that your hands are valuable grasping weapons; grasp any vital area with force. Use your fingernails to their best advantage for gripping any area.

4. TEETH

Bite into vulnerable areas whenever you have to; don't hesitate. It might mean serious injury or death for you if you do not. Bite the ear, the throat, or the areas which are grasped by the hand in illustrations 36 and 37.

5. HEAD BUTT (*See illustration 48*)

Drive your upper forehead into your opponent's face. If you are wearing a steel helmet, so much the better.

6. KNEELIFTS (*See illustrations 53 and 54*)

An opponent who grasps you in a frontal closed strangle hold is wide open for knee lifts to the groin and stomach. These blows will generally end the scrap right there.

7. NECK SNAP (*See illustration 114*)

Reach up behind your opponent's head, grasp his hair and snap down sharply to release yourself from this hold.

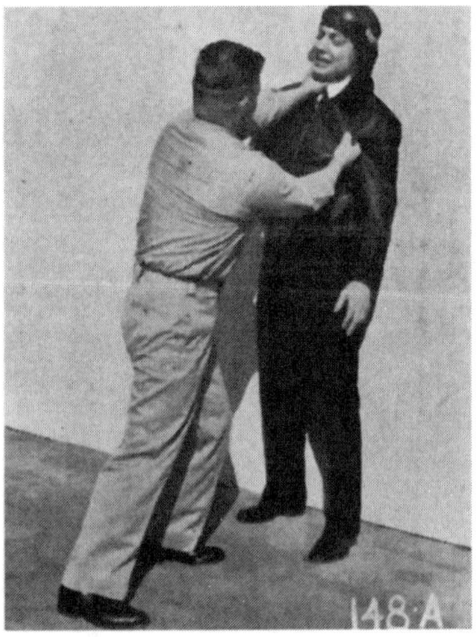

Reverse Wristlock

148A—Opponent attempts a one-armed strangle against a wall.

148B—Reach up over the clasping hand and again secure the little finger side, placing the other hand at opponent's elbow. Begin your turn out.

148C—Complete your turn, retaining your grasp on the wrist and elbow. Place your body between your opponent's side and arm, forcing his arm up; keep elbow locked. Opponent can be forced easily to deck or kicked in face.

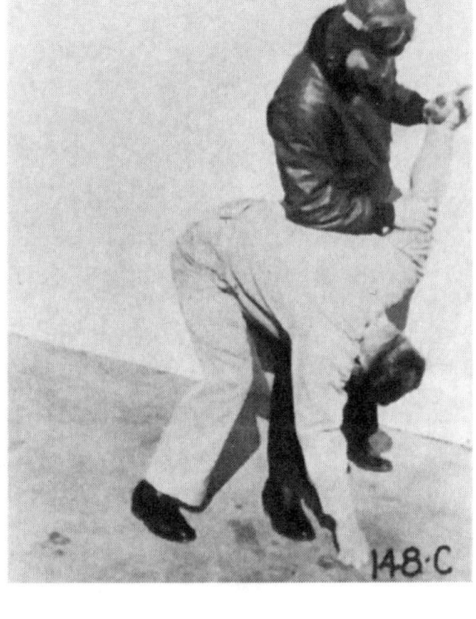

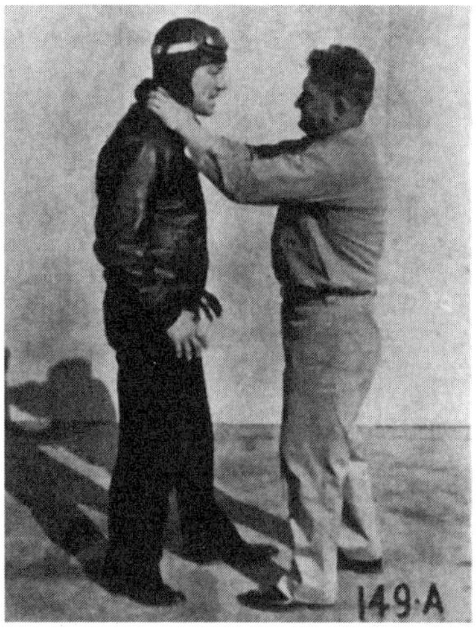

149-A

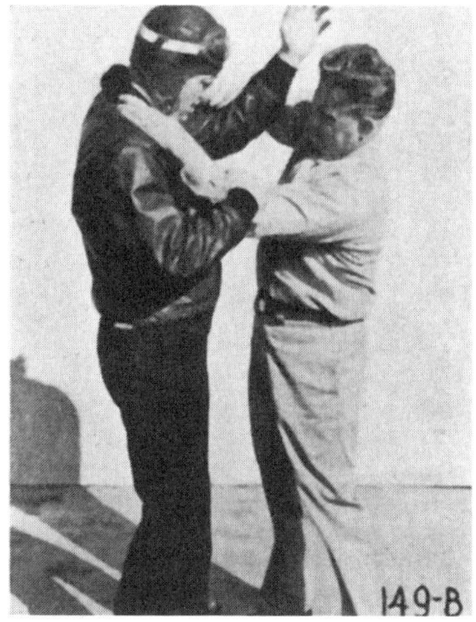

149-B

Elbowlock

149A—Opponent attempts an arms length strangle.

149B—Quickly bring the far arm up inside of opponent's and force his rear arm from throat.

149C—Continuing the movement of the far arm, bring it down outside opponent's am and secure with an elbow lock. Note that his arm goes under your armpit. Opponent's near arm is forced downward.

149D—Jerking quickly in and up bends opponent's elbow, causing a twisting elbowlock to be initiated. Use your near arm for additional leverage and kick to shin, knee, or groin. This pressure is very painful and the arm can be broken with added pressure.

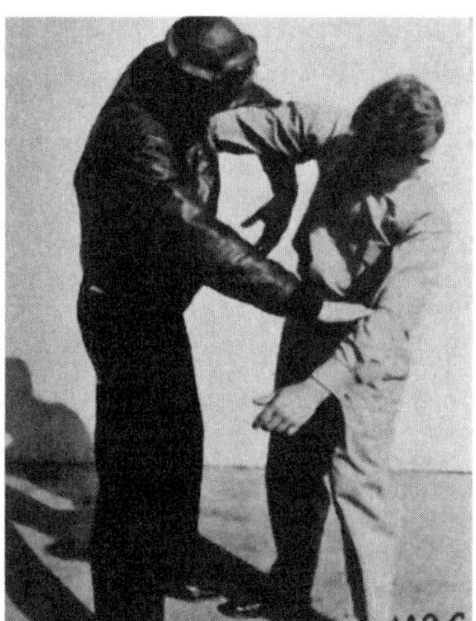

149-C

149-D

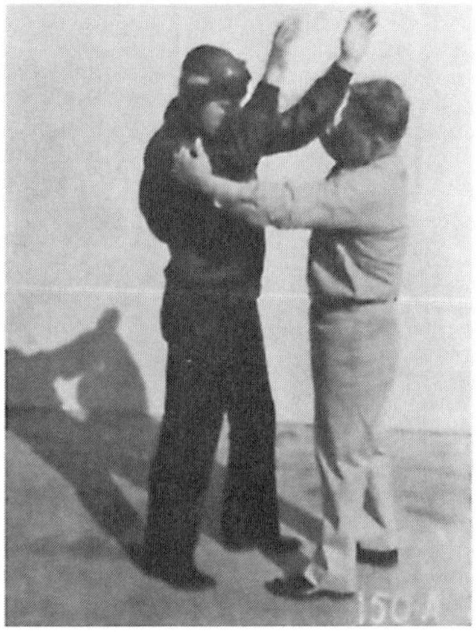

Strangle Hiplock

150A—Opponent has attempted an arms length strangle. The hold is broken by bringing the arms up inside those of opponent. This creates an outward pressure, breaking the grasp.
150B—Your far arm passes over and blocks opponent's arm. The near arm secures opponent at the elbow and pulls the arm across front of body.
150C—This causes opponent to spin a quarter turn and thus allows you to encircle his throat with the upraised far arm. Having obtained this strangle hold, step behind opponent's back.
150D—Locking your hands, use your hip as a lever, and pull opponent over it to the deck. As opponent strikes the deck, attack with knee drops, kicks, or hard blows.

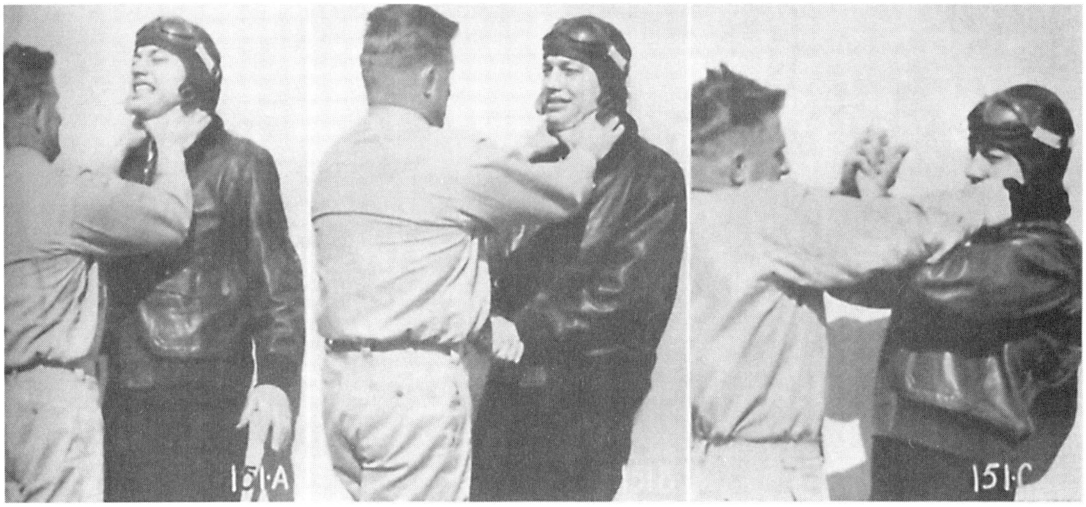

Arm Wedge

151A—When opponent has you pinned against a wall, the arm wedge is an effective weapon for counterattack.
151B—As opponent strangles you, bend your knees slightly and clasp your hands together, fingers not interlaced.
151C—With a quick upward thrust, this arm wedge is inserted between your opponent's arms causing his hands to release their grip on your throat.
151D—Continuing the action, bring your folded hands down across the bridge of your adversary's nose or to any vulnerable area of the face.
151E—Unlock your hands and strike with an edge of fist blow to jaw or to other areas within range.
151F—Striking out with the edge of the hand to the collar bone is also effective, as are other edge of hand blows to vulnerable areas. Note how the knee can be brought up to groin.

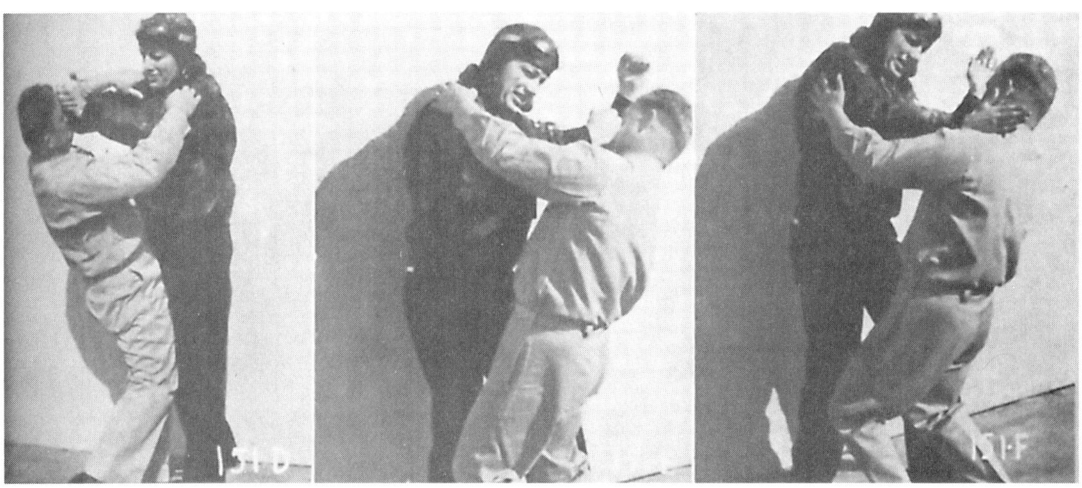

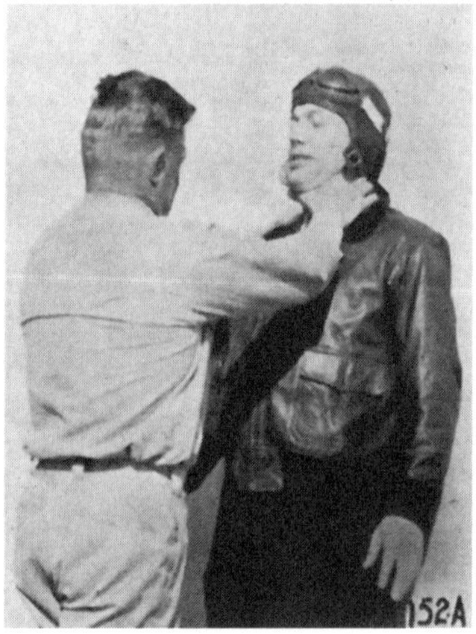
152A

152B

Other Escapes

152A—Opponent attempts a strangle and pushes you against a wall.

152B—By forcing upward against the locked elbows you can exert enough pressure to make your opponent release his grasp on your throat. This can be accomplished on one or both arms simultaneously.

152C—Or dig the thumbs or knuckles into the arm pits. Nerves in this area are very sensitive. (Note illustration 88)

152D—Or strike with the knife edge of the hand blow to the short ribs with all the force you can exert. The student can readily see how easy it would be to deliver many of the hand blows listed in the fundamentals from this position. The arm may be brought up inside your opponent's arms or you may strike from the outside.

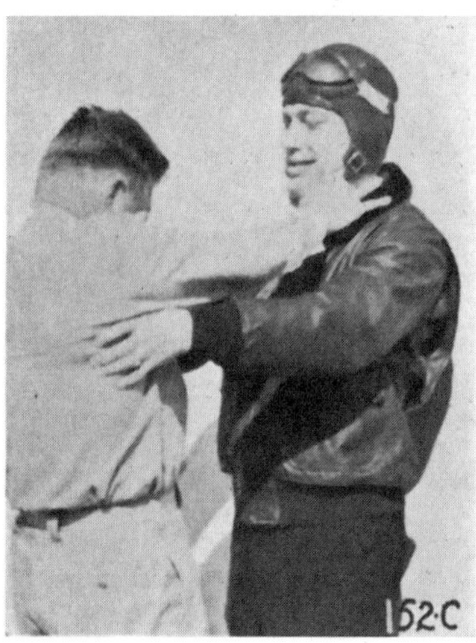
152C

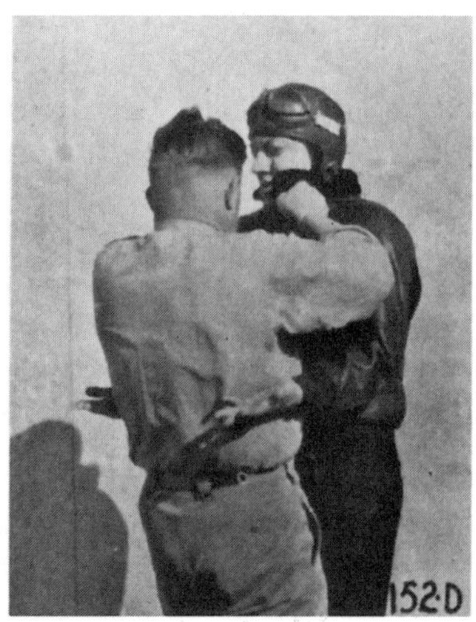
152D

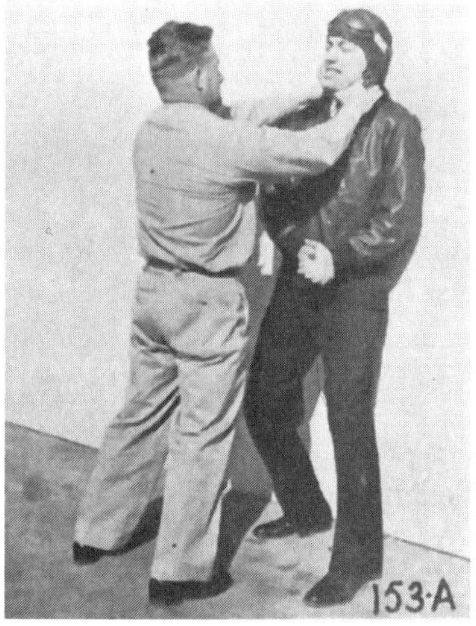

153-A

Outside Blow to Locked Elbow

153A—Opponent attempts to strangle; clasp your hands, but do not intertwine fingers, bend knees slightly. Notice that your arms are slightly to the side.

153B—Bring your arm up outside, against the locked elbow, using your other hand for added pressure and force. This blow will break the elbow or seriously injure it, due to the fact that your opponent's hand is locked against your neck.

153-B

E. Bodylocking

(Study maneuvers as listed referring to illustrations as needed.)

I. Arms Included

1. KICKS (*See illustrations 62, 63, 73*)

Again close quarters limit the number and desirability of kicking. Remember that the advantage is entirely with you when you are grasped in this fashion. An adversary unthinkingly leaves himself open for other more serious tactics.

2. HAND BLOWS (*See illustration 16*)

Close quarters and the fact that your arms are locked to your sides prevents the use of hand blows with the possible exception of a short edge of hand or edge of fist blow to the ribs or a short blow to groin.

3. GROIN GRASP (*See illustration 39*)

Although your arms are locked this does not prevent you from obtaining a grasp to the groin.

4. HEAD BUTT (*See illustration 48*)

Your forehead or helmet is an excellent weapon to the face of your opponent.

5. KNEELIFT (*See picture series 154*)

6. TEETH

Use your teeth to bite to vital areas. Bite especially to the areas grasped by the hand in illustrations 36 and 37.

II. Arms Free

1. KICKS (*See illustrations 62, 63, 73*)

Close quarters restricts kicking

2. HAND BLOWS (*See illustrations 4, 5, 7–10, 14–16, 19–26, 28, 32–35, and picture series 155*)

As your arms are free you can deliver a series of devastating blows to many vital areas about the head, throat, neck and ribs.

3. NERVE SHOCKS (*See illustrations 87–90*)

You are in position to use any of the mentioned nerve shocks. They may stand you in good stead, but don't depend too much on them. Use them and follow up with more serious maneuvers when your opponent flinches from the nerve pressures. However, if the strike to the nerve in the neck, shown in illustration 90, is done properly, you can easily overcome your adversary.

4. ELBOW BLOWS (*See illustrations 40–44*)

Such close quarters allows for excellent "elbow work." The elbows are capable of inflicting great damage to the face, shoulders and neck of an opponent.

5. KNEELIFT (*See illustrations 53 and 54*)

This position is ideal for the kneelift to the groin or stomach. Note, in illustration 155C above, the kneelift is used in conjunction with a hand blow.

6. NECK SNAP (*See picture series 156*)

7. LEG TRIP (*See picture series 157*)

8. EAR CLAP (*See picture series 235*)

Clapping both hands over the ears simultaneously, with force, may break the ear drums. Your opponent may drop as if he were struck by lightning, due to shock. However, when one is free to make such an attack other more effective maneuvers may be chosen.

9. REGULAR HIPLOCK (*See picture series 145*)

The regular hiplock works from this position, but you must first force your opponent away from you so that you can step through to use your hip as a lever.

10. REVERSE HIPLOCK (*See picture series 146*)

Again you must force your opponent away so that you can step through and use your hip to throw. This may be done by using a kneelift or forcing your thumbs into and under the short ribs.

11. DOUBLE WRISTLOCK (*See picture series 143*)

The double wristlock, an extremely valuable tactic, can be worked whenever an opponent's arm and wrist are within grasp. In order to secure the wristlock from this position you must first force yourself away from your opponent. It also necessitates reaching much higher than that illustrated in series 143. However, once you have secured your double wristlock carry on with your throw, or any other attack you deem desirable.

12. ELBOWLOCKS (*See illustration 104–106*)

All that need be done to get an elbowlock on an opponent is to grapevine the arm and snap it inward. Locking an opponent securely enables you to attack him very readily. Single elbowlocks may be obtained on both of opponent's arms simultaneously.

Remember, the above maneuvers all deal with frontal attack. Some can be used in various situations, others cannot. The fact that a particular maneuver was listed alone, does not necessarily mean that it should be worked alone anymore than a particular wrestling tactic would be in a regular match. The techniques and tactics are all fitted together and one leads to the other. If one action does not work at the moment, do not become discouraged. Try another. But above all, you must make up your mind what you are going to do and then do it by acting quickly.

Kneelift

154A—Opponent has you grasped in a bodylock with your arms included.
154B—The knee is lifted into the exposed crotch with force. Notice the vulnerability of this position, and learn to keep out of it yourself.

Hand Blows

155A—Opponent has bodylocked from the front. Note his vulnerability to attack
155B—Bring up your arm, jab viciously to opponent's eyes, gouge them thoroughly until he releases you.
155C—The illustration shows the effect of a heel of hand blow to the chin accompanied with a kneelift to the groin. Generally, either one will suffice to subdue your opponent.

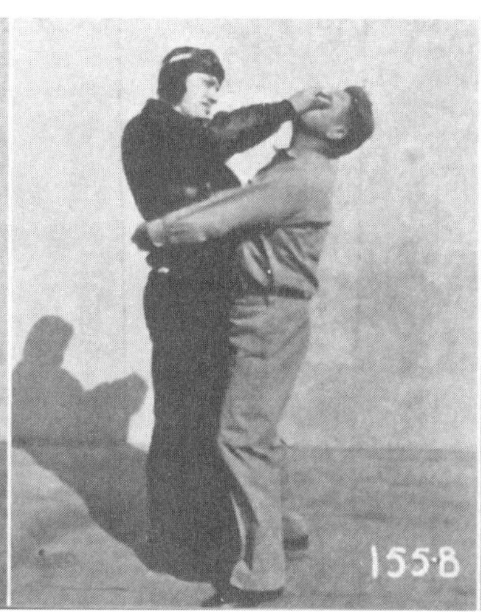

Neck Snap

156A—Opponent has secured you in a front bodylock, arms free.

156B—Reach up behind opponent's head with either arm and get a good grip on his hair.

156C—Holding the hair firmly snap the head back sharply to the rear. This quick snap will generally injure the neck.

156D—If your opponent still has some fight left in him, follow up the neck snap with a knuckle jab to the larynx. Strike hard and you can kill your man.

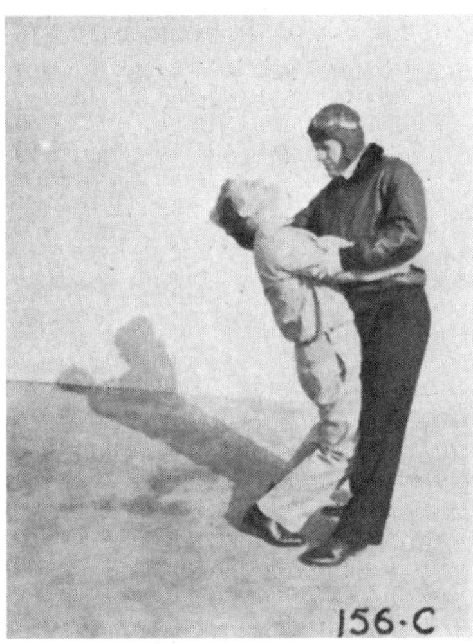

Leg Trip

157A—Opponent obtains a frontal bodylock, and applies a "bear hug" in order to double you backward.

157B—Counter the "bear hug" by one of your own and throw one leg and foot behind a leg of your opponent and trip backward to the deck and apply a knee drop to the groin or stomach with your other leg (see illustrations 59 and 60), or kick his body in vulnerable areas.

CHAPTER 7

Kicking Maneuvers

WHEN A GOOD DEFENSE is set up against arm and body tactics, kicking may be resorted to. Occasionally, too, a situation arises where one loses his balance, and is thrown or knocked to the deck. This predicament, though serious, is not fatal. There are many offensive maneuvers which can be carried out from this position.

If knocked to the deck, never attempt to get up regardless of the position of your assailant. He has many opportunities to kick or attack many vulnerable parts of the body. Your chances are best, for the moment at least, if you keep your feet toward assailant. Since the legs are in position, use them to parry or ward off blows or kicks. The legs and feet offer a maximum defense and can be used offensively if adversary approaches within range.

Arise from the deck with the feet toward opponent and in case of a sudden attack, counter with leg kicks. The legs are not so vulnerable to attack, and kicks or blows to them would be less severe than blows directed at the arms, head or body.

This chapter, divided into two parts, will discuss first, standing or upright maneuvers and, second, defensive and offensive maneuvers from the deck.

A. Upright

1. AVOIDING KICKS

A kick may be avoided by springing away, jumping to one side, or rushing your opponent so that the kick has little or no effect. Also, it is possible to use your forearm as a shield to block the kick. In this tactic your forearm blocks the kick by placing it against the shin. For another method of avoiding kicks see picture series 158.

2. COUNTER KICKS AND TRIPS (*See picture series 159*)

3. CROSS ARM CATCH (*See picture series 160*)

4. STEP INSIDE—FRONTAL ATTACK (*See picture series 161*)

5. GRASP LEG—TOE HOLD INWARD—BAR TOE—ATTACK FROM REAR
(*See picture series 162*)

Avoiding Kicks

158A—Do not give yourself away before your opponent starts his kick. Stand firm and alert.
158B—As opponent kicks, step to one side, that is, throw one leg back and turn to side keeping one fo
in place.
158C—This is another view of illustration 158B. Note that the body is turned away, but that one fo
remains in place. This allows for counterattack on your part. You protect your groin, but if you d
misjudge, the side of your stationary leg will get the kick. This is better than catching a kick in a vital area

Counter Kicks and Trips

159A—It will be noted that this is a continuation of picture series 158 above. As opponent kicks, his le
is grasped, one hand over leg at calf and the other over toes.
159B—From this position a counter-kick to the knee cap, shin, or groin can be delivered.
159C—Or by raising the leg with the hand and arm that originally grasped the kicking toe, step behir
opponent's foundation leg and trip him to the deck on back or head. As he falls backward, he has r
means of catching himself or breaking the fall. Raise the upheld leg as you trip for additional leverage.

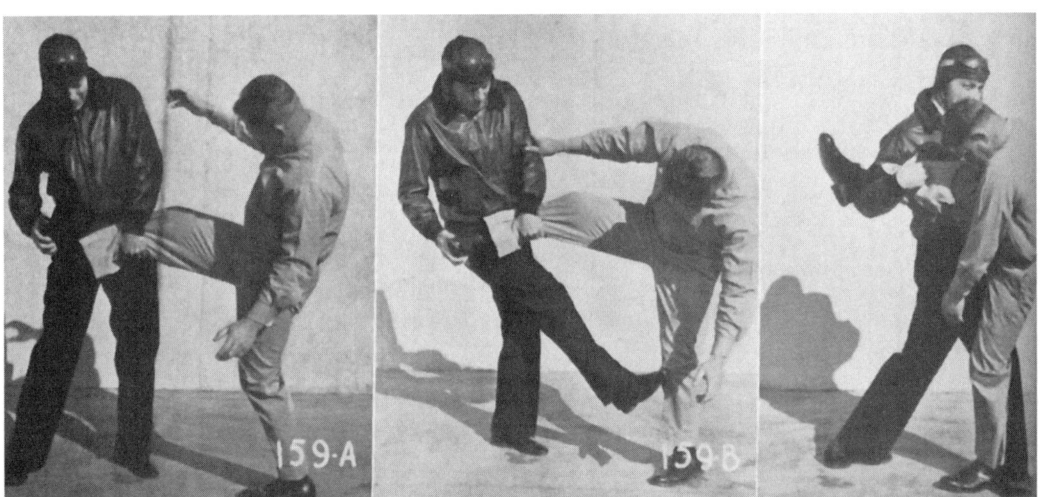

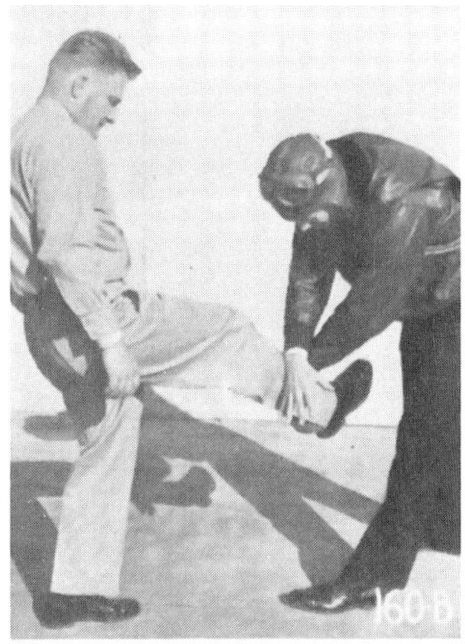

Cross Arm Catch

160A—As your opponent starts his kick, do not retreat, step in and get your arms in position in front of you.

160B—Cross your arms and catch the kick in the "V" made by your crossed arms.

160C—Use your hands, grasp opponent's kicking leg and raise it higher than he intended it to go.

160D—Retain your grasp on opponent's leg, and kick to the groin. You may also kick to opposite leg or stomach.

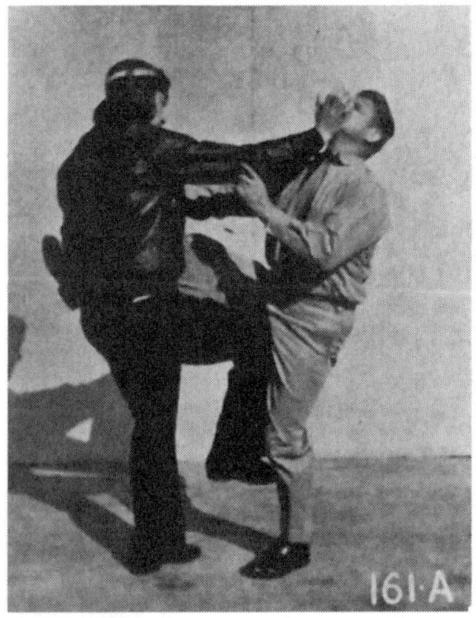

Step Inside—Frontal Attack

161A—As opponent kicks, step inside the leg and grasp it with one arm. Hold it firmly at your side. The counterattack is a heel of hand blow to the chin and kneelift to the groin. Other forms of frontal attack may also be delivered. It is also possible to step behind your opponent's foundation leg and trip him to the deck for further attack.

Grasp Leg—Toe Hold Inward—Bar Toe—Attack from Rear

162A—As opponent kicks, turn away, as shown in picture series 158 above and grasp the kicking leg, one hand over calf and other over toes.

162B—By quickly twisting the toe inward opponent must turn his back to you or receive a severely injured ankle.

162·C

162·D

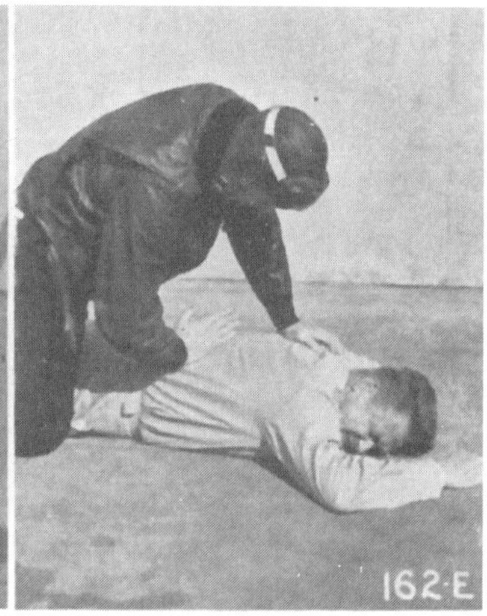

162·E

Grasp Leg—Toe Hold Inward—Bar Toe—Attack from Rear (continued)

162C—A shove forward will force opponent to the deck; however, this is not necessary as the pressure created by your stepping over the held leg at the hock of the knee will force him to the deck.

162D—Keeping the leg in the hock of your opponent's knee, go to your own knees and force your opponent's leg back with your upper leg at the crotch. Notice that his toes are protruding just over the hip at the crotch. In other words your opponent is held firmly to the deck with a bar toe hold and additional pressure at the ankle may dislocate or break the knee.

162E—Having forced an adversary to the deck and having him clamped down with the bar toe hold (note position of opponent's foot), you are free to attack his back with various hand blows, one being the edge of hand blow to the kidneys as shown here.

162F—Or reach over the head, clamp fingers in eye sockets and snap head back, injuring the neck.

162G—Or use both hands, one to back of head, other under chin and snap head sideways, again injuring neck.

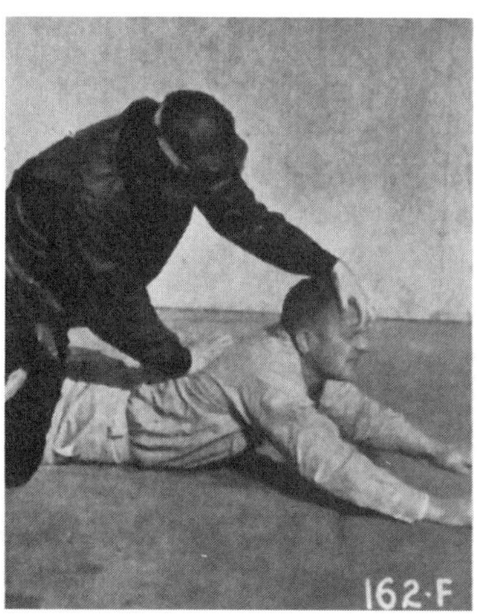

162·F

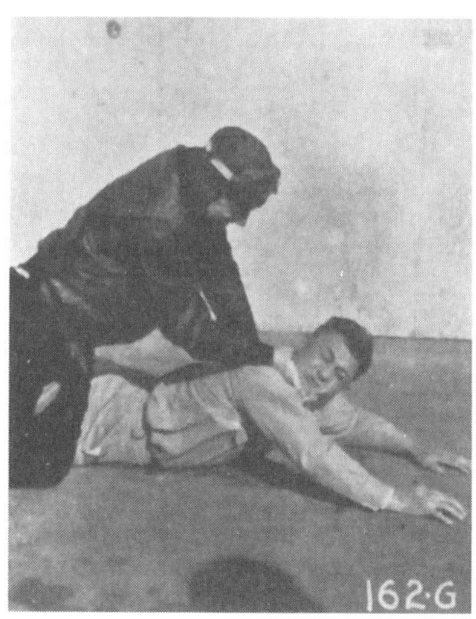

162·G

B. From the Deck

1. TOE HOOK—KNEE KICK (*See picture series 163*)

2. SCISSOR KICKS TO BAR TOE HOLD
    a. Outside Kick (*See picture series 164*)
    b. Inside Kick (*See picture series 165*)

3. GRASP KICKING LEG—GROIN KICK (*See picture series 166*)

4. LOW BODY BLOCK (*See picture series 167*)

5. MULE KICK (*See picture series 168*)

6. ROLL OUT OF RANGE—COUNTERKICK

If a situation arises where time does not permit you to get your legs in position, quickly roll away, spin and kick out at your opponent until you can get up or into position for attack. Using your legs to kick with, often will discourage an opponent from approaching too close, thus giving you time to get on your feet.

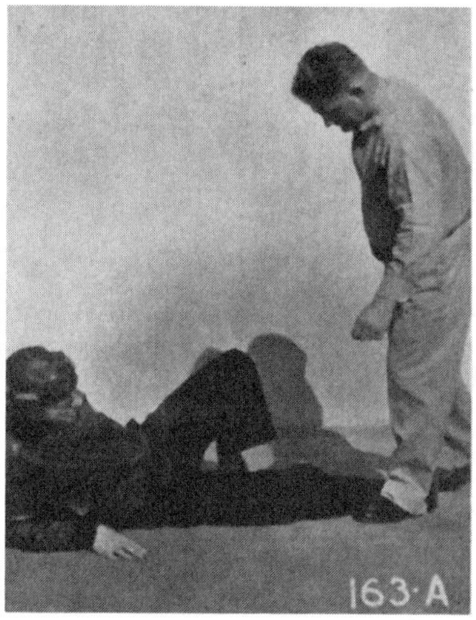

Toe Hook—Knee Kick

163A—When you are knocked to the deck, always try, if possible, to keep your feet toward your opponent. Lock toe of your lower leg behind opponent's heel of balance leg.

163B—Now kick hard to opponent's knee with your upper foot. This kick will cause a severe injury to the knee since opponent is moving forward to attempt a kick. Come up to your own feet and counterattack.

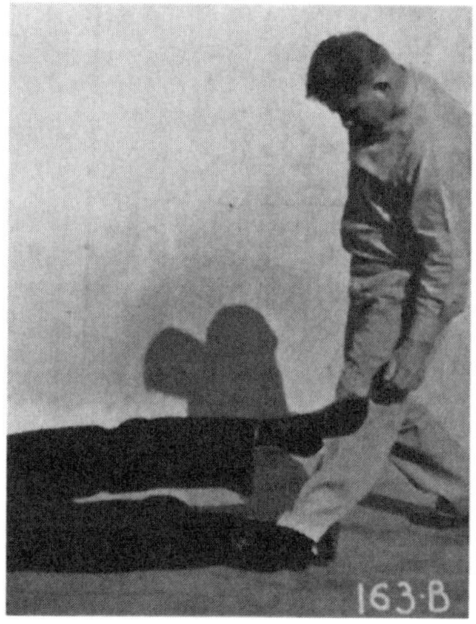

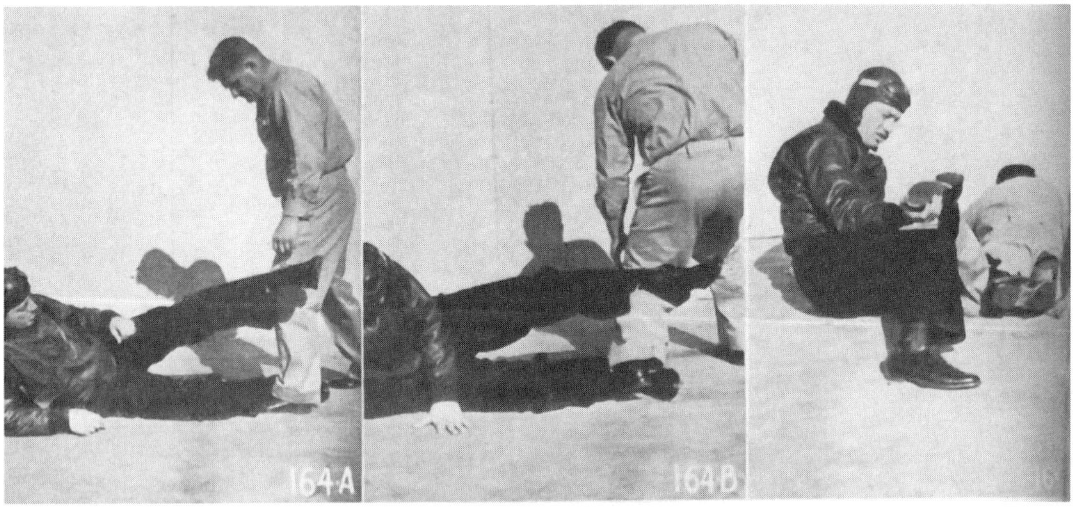

Scissors Kick—Outside

164A—Again hook opponent's leg with toe on the inside. Now kick with your upper leg against th outside of opponent's balance leg.

164B—The kick is delivered with force to the knee area of opponent's leg, thus causing him to bend kne away to prevent serious injury. Notice the scissor motion to your legs, they are moving in opposi directions, the lower leg comes forward while the upper leg moves backward.

164C—Continued pressure throws opponent to the deck. The leg that struck at opponent's knee remair in place while you grasp his toes and begin an upward pull. The bar toe hold is now taking form.

164D—Pulling up on opponent's toes, your body is brought against the toe and ankle. Your under leg still in the hock of opponent's knee.

164E—The final result is the position illustrated above and allows you use of your hands to attac opponent's back, neck and head as heretofore described. (Note how your leg acts as a bar or lever o opponent's leg.)

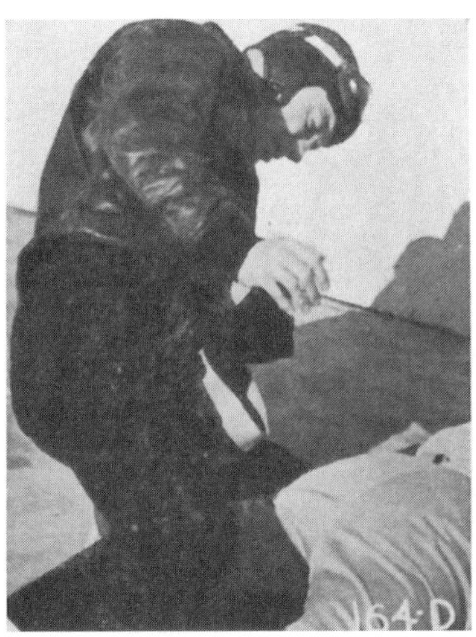

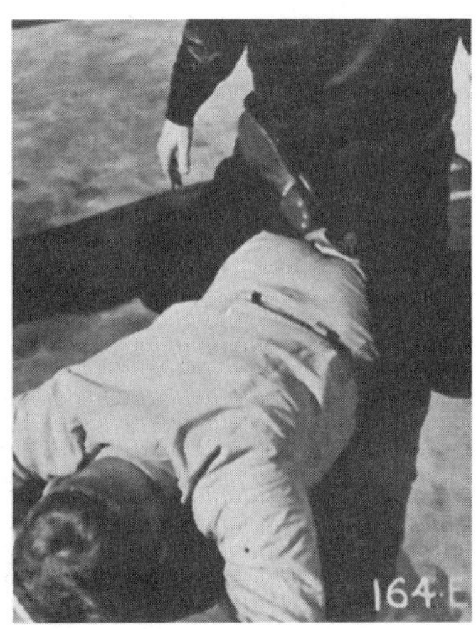

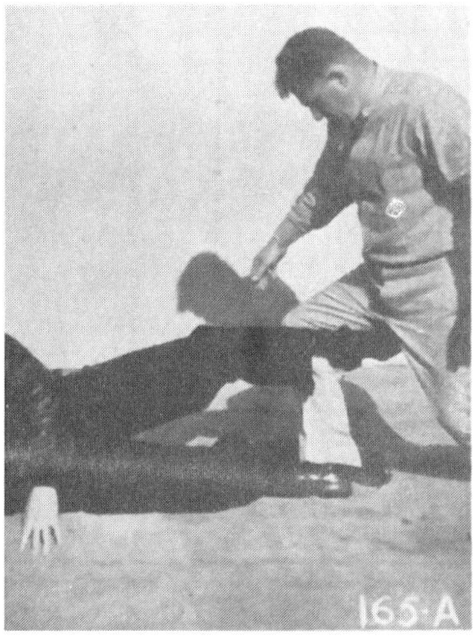

Scissors Kick—Inside

165A—Opponent's leg is hooked from the outside by placing your toes behind the heel of his balance or foundation leg. The upper leg is kicked against the inside of opponent's knee with force. This scissor kick causes opponent to fall to his back on deck.
165B—By reaching down and grasping your opponent's ankle from the outside (as shown) and leaving your upper kicking leg at the side of opponent's knee. jerk up quickly. This pressure will dislocate the knee.

Grasp Kicking Leg—Groin Kick

166A—(Note that opponent is in an ideal position for the toe hook-knee kick.) Opponent attempts to kick you in the body. Try to keep feet toward opponent at all times, if possible.
166B—Use forearm to block kick and seize ankle with other hand.
166C—Retaliate with a heel kick to opponent's groin as shown, or strike to stomach, chest or face.

167-A

167-B

Low Body Block

167A—If you are caught on your hands and knees and opponent is attempting a kick to the body or face.

167B—Drive your body against opponent's kicking leg at the knee and grasp ankle with your hand, using arm to maintain grasp securely.

167C—This will cause opponent to fall backward to the deck because of the pressure against the locked knee. Your other hand now secures the ankle from the outside. Other hand holds firm.

167D—As you body rolls to the inside of opponent's locked leg at the knee which is now under your shoulder, quickly jerking upward on the ankle will break or dislocate the knee. This causes a very serious injury and caution should be used in practice.

167-C

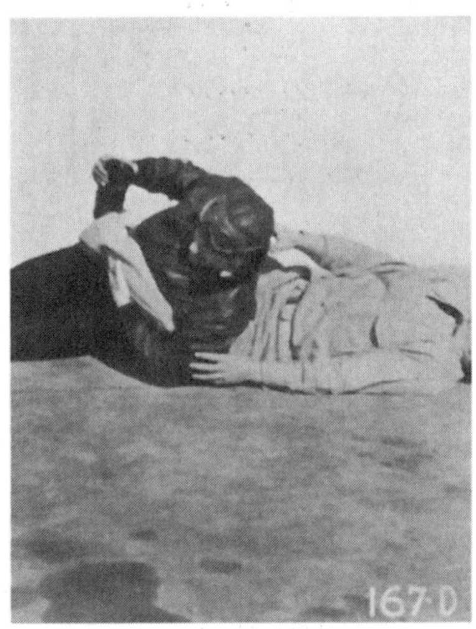

167-D

168·A

168·B

Mule Kick

168A—You have been knocked or have fallen to the deck. Opponent will more than likely attempt a kick.

168B—Turn away from opponent but keep your eyes on and your feet toward him.

168C—As opponent approaches or attempts to kick, bring both feet up with a sharp double kick to the chest, face, groin or stomach. Put all the strength of your legs behind the blow and you can knock your opponent senseless to the deck.

168D—If by chance, or luck, opponent grasps one of your kicking legs, kick again with your free leg to vulnerable areas.

168·C

168·D

CHAPTER 8

Rear Attack

CIRCUMSTANCES and conditions may be such as to give an assailant the opportunity to attack from the rear. Again it will be found that the techniques and maneuvers are given in the manner that they are most likely to occur; first, pushing and striking; second, grasping; third, strangling; fourth, bodylocking.

It is a natural reaction when attacked from the rear to become excited, and struggle wildly. The chances are still good of coming out of such an encounter unscathed if the following principles are kept in mind. First, block, if possible, any maneuver applied by the assailant, and second, either by force or feinting movements throw the assailant off balance and guard, and counterattack any openings or vital areas with an unsurpassed viciousness.

A. Striking or Pushing

(Study maneuvers as listed referring to illustrations as needed.)

1. EDGE OF HAND OR FIST BLOWS (*See picture series 169*)

2. HEEL KICKS TO BODY (*See picture series 170*)

3. SINGLE OR DOUBLE MULE KICK (*See picture series 171*)

4. TURN AND HIPLOCK (*See picture series 145 and 146*)

In order to work this maneuver you must be capable of catching your balance quickly and turning instantly into your adversary to meet his rush. If this can be accomplished, your maneuver is then worked exactly as shown in picture series 145 or 146.

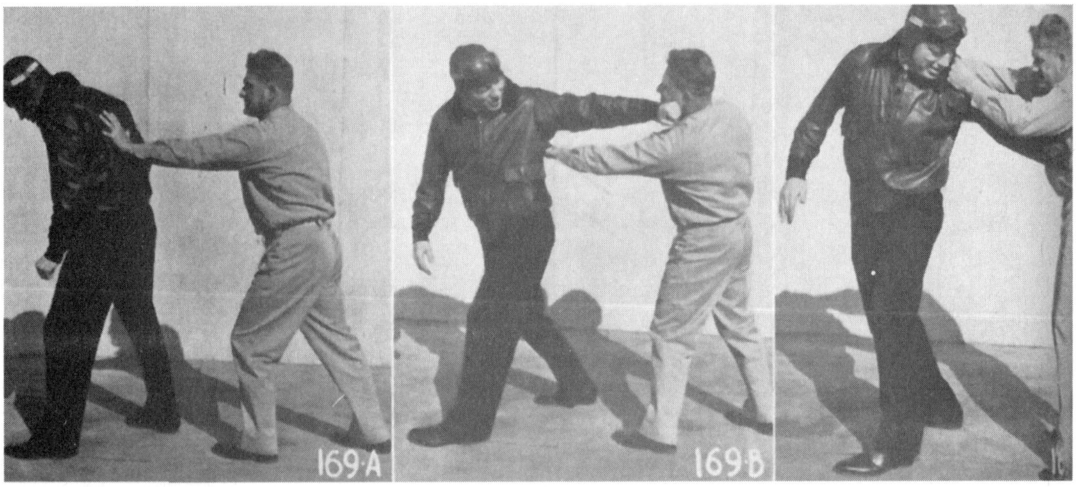

169·A 169·B

Edge of Hand or Fist Blows

169A—Opponent comes up behind and gives you a shove forward.

169B—If the shove forward does not throw you to the deck, counter backward with an edge of fist blow to the jaw (as shown) or to the ribs.

169C—An edge of hand blow to the ribs is also effective. The same blow can be delivered to opponent's face. These blows can be delivered with either hand, depending on your balance and turn.

Heel Kicks to Body

170A—Opponent pushes you forward.

170B—Bring your heel up into opponent's groin (as shown), his shin or knee-cap. This is also practical you are pushed to your hands and knees.

170·A

170·B

Single or Double Mule Kick

171A—Opponent pushes you forward and off balance; you go to your hands.

171B—From this position you start a kick up to your opponent's body.

171C—The kick can be delivered to the face, chest (as shown) or groin with either one or both feet. (Note picture series 168)

B. Grasping

(Study maneuvers as listed referring to illustrations as needed.)

1. KICKS (*See illustrations 74–77*)

Grasping does not mean bodylocking. Your opponent has a grip on your clothes or arms. Using heel kicks to vulnerable areas will make your escape easier.

2. HAND BLOWS

Hand blows, such as shown in picture series 169 above, can be used against a grasping opponent.

3. ELBOW BLOWS (*See illustrations 45–47*)

It is possible to swing over or under the grasping hands and arms of an opponent and strike him in the face, solar plexus, or groin with an elbow brought back with great force. It affords an effective and efficient weapon if used properly.

4. DOUBLE ELBOWLOCK (*See picture series 172*)

5. TWISTING ELBOWLOCK (*See picture series 173*)

6. REGULAR WRISTLOCK (HAIR GRASP) (*See picture series 135 and 136*)

The regular wristlocks are worked in the same manner, but the technique of getting the hand from the hair is slightly different. It will be recalled that in applying a regular wristlock from in front you reach straight across or over the back of the hand to the thumb side. You do this as you reach back for your grip. Now turning out you will discover that your hand position is reversed, but don't let this confuse you. Carry on through the maneuver regardless of hand position.

7. REVERSE WRISTLOCK (HAIR GRASP) (*See picture series 137*)

In this tactic you reach over the back of hand to little finger side. Don't let the fact that your hands are reversed deter you from carrying on through with the reverse wristlock as shown previously.

8. FINGERLOCKS

The various forms of fingerlocks are all available to you when grasped by the hair from behind. With slight variations to fit your own needs, you can bring your opponent to the deck without much difficulty.

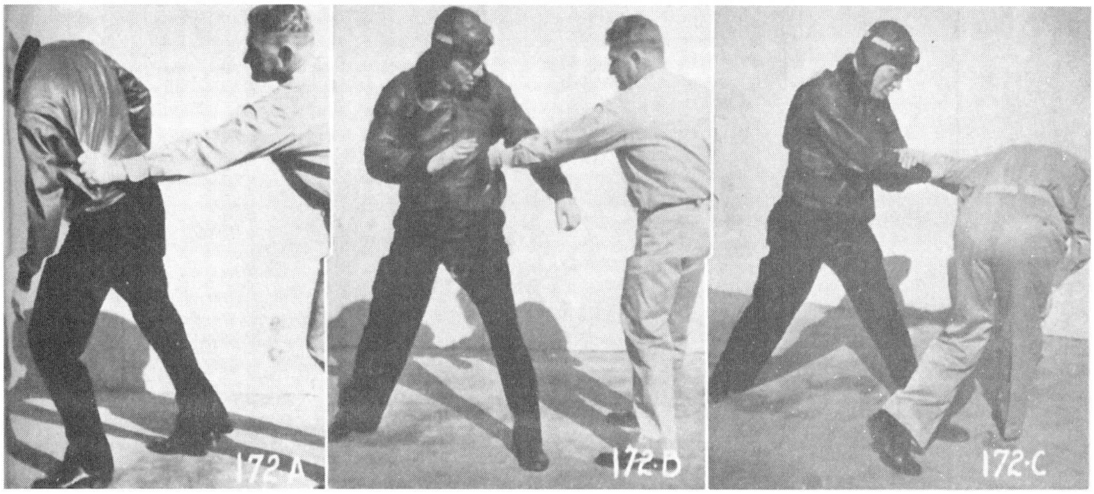

Double Elbowlock

172A—Opponent is grasping clothing from the rear, approximately at the waist.
172B—Turn outside opponent's arm and throw your arm over it. Your other hand is also brought up and—
172C—Placed on opponent's upper arm while your hand secures itself on the forearm. The pressure on the upper arm keeps elbow straight. The elbowlock is on the elbow from the outside. If you had turned inside the grasping arm the elbowlock pictured in illustration 105 could be applied.

Twisting Elbowlock

173A—Opponent grasps clothing at the shoulder. As this would make it difficult to apply a double elbowlock, do the following.
173B—Turn outside opponent's grasping arm and throw your arm *under*
173C—And bring the forearm over into the bend of the elbow. Secure your hands together and bend your body forward. Thus a twisting elbowlock is applied. Stepping behind opponent an additional pressure on arm will force him to the deck where you can attack further with knees or feet if you wish.

C. Strangling

(Study maneuvers as listed referring to illustrations as needed.)

I. Arm's Length

1. KICKS (*See illustrations 74–77*)

Although not bodylocked, you can still place well-directed heel kicks to vulnerable areas of your opponent's body.

2. HAND BLOWS (*See illustrations 17, 25, 28, 29 and picture series 169*)

Opponent need not be directly behind to deliver these hand blows to vulnerable areas. Get the habit of using them with force.

3. ELBOW BLOWS (*See illustrations 45–47*)

An opponent may, at any time, leave himself open to elbow blows to these vital areas.

4. DUCK HEAD—TURN (*See picture series 174*)

5. GRASP THUMB—DUCK AND TURN—LOCK ELBOWS

Reaching up behind neck with both hands, secure both thumbs, duck and turn as shown in picture series 174. Maintaining your grip on the thumbs, you can now lock one arm against the other, and with a downward pressure, cause opponent to bend over where you can kneelift to his face. (This maneuver is well illustrated in the U. S. Navy Training Film "Hand-to-Hand Combat.")

6. GRASP THUMB—TURN PALM UP—DRIVE SHOULDER TO ELBOW (*See picture series 175*)

7. TURN OUTWARD AGAINST ELBOW

If you turn out quickly to either elbow, striking it with an upraised arm, you will succeed in breaking your opponent's grasp on the neck. Also it is possible to use the tactic described and illustrated under picture series 173.

II. Closed Strangle

1. KICKS (*See illustrations 74–77*)

All the heel kicks can be used freely when an opponent attempts a closed strangle. Range is short.

2. HAND BLOWS (*See illustrations 17, 29*)

Hand blows are limited in this form of attack. It is also necessary to use the hands to keep breathing. However, use them to the groin whenever possible. A sharp strike to the groin will generally get you out of any tight spot.

3. GROIN GRASP (*See illustration 39*)

Need any more be said!

4. ELBOW BLOWS (*See illustration 46 and 47*)

These two elbow blows can cause serious damage to an opponent, so use them!

5. HEAD BUTT (*See illustration 49*)

If you have a steel helmet on, so much the better!

6. FLYING MARES (*See illustrations 128 and 129*)

- a. Using Knee (*See picture series 176*)
- b. Using Hips (*See picture series 177*)
- c. Using Knee and Hair (*See picture 178*)

7. GRASP WRIST—PUSH UP ON ELBOW TO HAMMERLOCK (*See illustration 108*)

This maneuver is illustrated in any swimming or life saving manual. Reach up with one hand and grasp the wrist; the other hand pushes up on the elbow forcing it over head. Maintaining grasp on arm, go behind opponent for a simple hammerlock.

8. TWIST AND SIDESTEP—TRIP OR LIFT

- a. Trip (*See picture series 179*)
- b. Lift (*See picture series 180*)

9. REVERSE WRISTLOCK (*See picture series 181*)

10. REVERSE DOUBLE WRISTLOCK (*See picture series 182*)

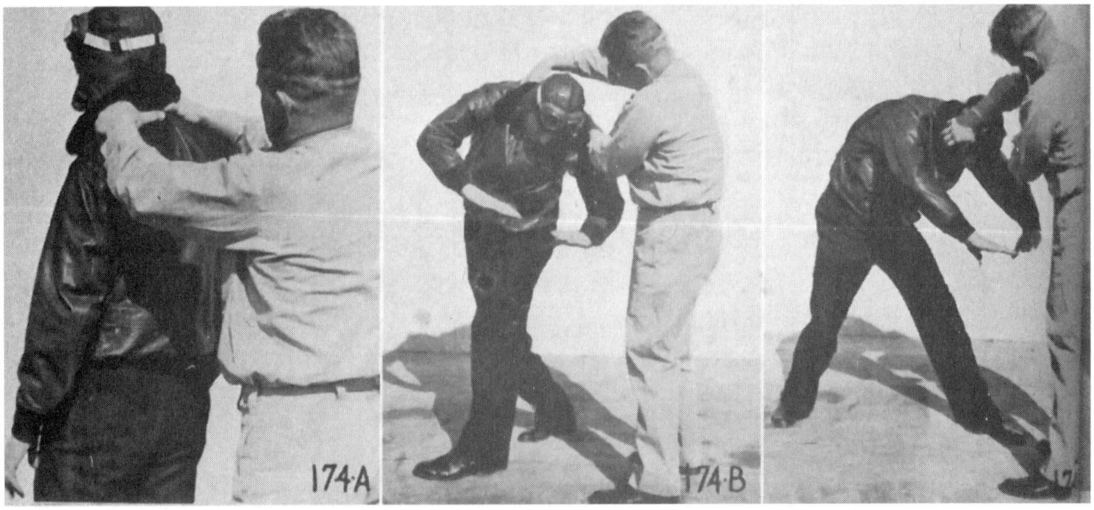

Duck Head—Turn

174A—Opponent has secured an arms length strangle or grasps clothing.

174B—Ducking the head and turning quickly will release you from an opponent's grasp regardless of h strength. The arms are out in front of body to ward off any kicks or kneelifts to your face or body.

174C—Continuing the turn, twists opponent's arms, and leaves you free to step back or close in fo further attack. The hands are still used to ward off blows or kicks.

Grasp Thumb—Turn Palm Up—Drive Shoulder to Elbow

175A—Opponent is attempting an arms length strangle from the rear. Reach back and over hand at nec and secure the thumb side of one or both hands.

175B—Prying opponent's hand loose, turn palm up and then use other hand for additional leverag straighten opponent's arm over shoulder.

175C—With arm straight and elbow locked, drive shoulder up into the arm. This will either dislocate o break opponent's arm. Note that the arm is held with both hands. From this position a flying mare ca be used to throw opponent to deck. Bend at hips and pull sharply down on arm. He'll go over muc easier than you imagine. Don't attempt this flying mare in practice, it's too dangerous.

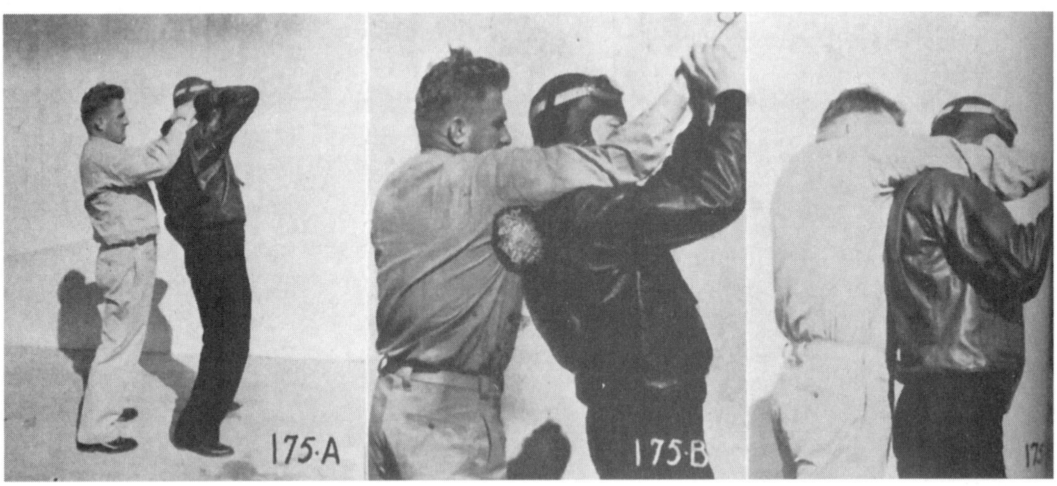

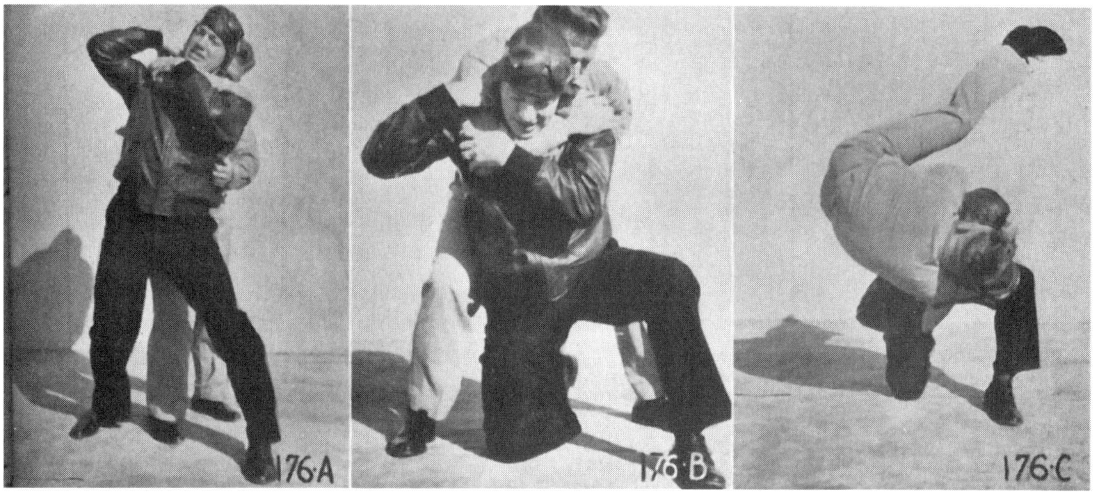

lying Mare Using Knee

76A—Opponent attempts a closed strangle from the rear. Reach up and secure upper arm with one and and elbow with other.

76B—Drop to knee, on side that opponent's arm is secured. Note particularly position of your hands on pponent's arm.

76C—Bending forward quickly, snap opponent over shoulder to the deck. Under actual combat onditions, the drop to the knee and the bend forward would be more or less simultaneously done.

lying Mare—Using Hips

77A—With palm up and elbow locked you are in position to throw opponent.

77B—(See also picture series 175) Bending quickly forward at the hips pull opponent over head to deck. Opponent's arm was shifted in the throw to prevent injury.) This is a dangerous maneuver, as the arm is ften broken when the elbow is locked. In this flying mare note that the legs are spread for balance as ictim comes over back. Both arms can be locked over the shoulders if they can be secured.

lying Mare Using Knee and Hair

78A—If opponent is unwary enough to place head in a position for you to grasp his hair, do so and rop to deck on knee to side which hair is held. Snap forward and pull opponent over shoulder to deck.

Twist and Sidestep—Trip

179A—Opponent has applied a closed strangle. Your hands immediately go up to keep opponent's locking arm from cutting off your wind. In other words, your hands prevent a full strangle from being applied immediately.

179B—Quickly shove your hips out to side still keeping your grip on opponent's arm. Note that your leg has taken a side step.

179C—You have changed opponent's maneuver to a simple headlock. Reach behind him with the inside arm and secure opponent's far arm and place one leg behind him.

179D—Pull opponent back over leg to the deck. Drop him on his head.

Twist and Sidestep—Lift

180A—From the position previously obtained in illustration 179C, you are in position to lift your opponent from the deck. Your far arm locks opponent's far arm (arrow) while your near arm is inserted in opponent's crotch from the rear.

180B—Raise opponent from deck with the grasp just secured.

180C—Release your grip on opponent's arm, maintaining it on his legs, drop him to the deck on his head with force and attack as you see fit with knees or feet.

Reverse Wristlock

181A—Opponent has secured a closed strangle. Reach up and over opponent's hand and secure the little finger side. Your other hand is placed on the elbow of the same arm.

181B—Push arm outward and turn little finger side of opponent's hand up.

181C—Turn under opponent's held arm maintaining grip with both hands.

181D—Step away from opponent and force his head down by pressure on the elbow. Notice that the position of your hands on opponent's arm has not changed.

181E—Toe kick to opponent's face (as shown), neck or chest. Great damage can be wrought. The arm may be broken by pulling up on the wrist and pushing down on the elbow.

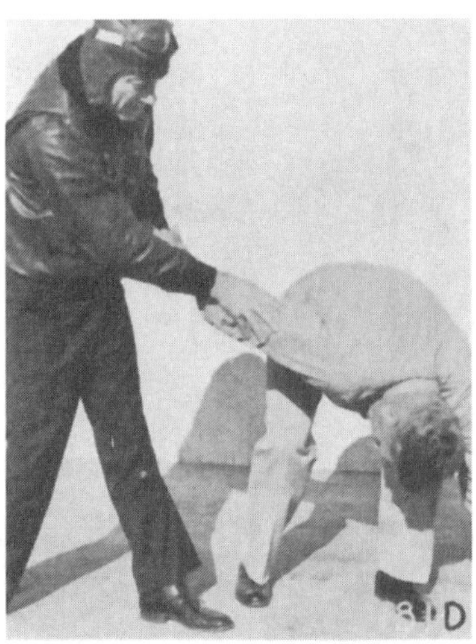

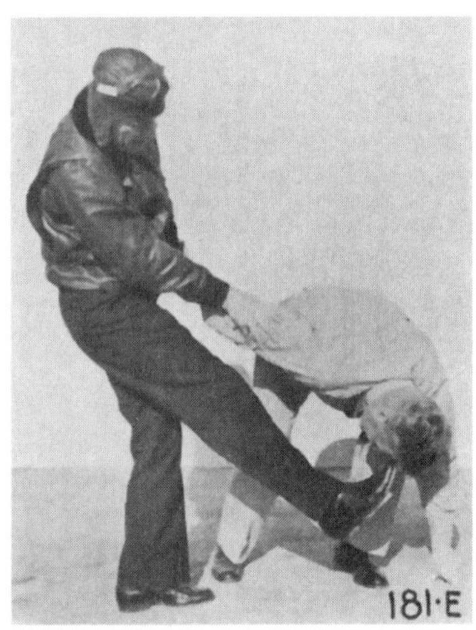

everse Double Wristlock

32A—Opponent has a closed strangle. Reach up under opponent's hand and grasp it so that your ngers lie against his palm and thumb is over back of wrist. Your other hand is placed just above the bow of opponent's same arm.

32B—Force opponent's arm outward and up, maintaining grip.

32C—Turn toward your man.

32D—Still retaining your grip on the wrist, bring your other hand up so that it locks over your own rist. Now press opponent backward. Continued pressure will break the arm or force him to the deck. ou can use your inside leg to trip if you desire. Study illustration 103 for a close-up of the double reverse ristlock. Observe that your arm passes behind opponent's upper arm, comes through at the bend of e elbow, and locks over your own wrist.

D. Bodylocking

(Study maneuvers as listed referring to illustrations as needed.)

I. Arms Included

1. KICKS (*See illustrations 74–77*)
These kicks are practicable against an opponent who bodylocks his victim from the rear.

2. HEAD BUTTS (*See illustration 49*)
The harder your head, the harder you can strike. A steel helmet is always practicable.

3. ELBOW BLOWS (*See illustrations 46 and 47*)
As your arms are pinioned to your sides, you must squat and raise your arms to get room to deliver the blows.

4. GROIN GRASP (*See illustration 39*)
Arms are pinioned, but you can still execute this maneuver.

5. FINGERLOCKS (*See illustrations 97 and 98*)
Although your arms are locked to your sides, you can still get free by using a fingerlock, but with force.

6. SQUAT—RAISE ARMS—TURN IN—GO BEHIND—INSIDE BAR HAMMERLOCK (*See picture series 183*)

II. Arms Free

1. KICKS (*See illustrations 74–77*)
Use kicks whenever you can benefit from the results obtained.

2. HEAD BUTT (*See illustration 49*)
The nose is the best objective.

3. NERVE SHOCKS (*See illustrations 83–85*)
Since your opponent's hands are in position, these nerve shocks are easily applied. You will discover that the "Knuckle jab" to the back of the hand is ideal for counterattack.

4. ELBOW BLOW (*See illustration 45*)
Can be delivered to the face and head from one or both sides.

5. STRIKES OR GROIN GRASP (*See illustrations 17, 29, 39*)
This grasp or blows are effective and cause serious injury.

6. FINGERLOCKS (*See illustrations 97 and 98*)
With both arms free you can break any bodylocking grip with these fingerlocks.

7. DOUBLE WRISTLOCK (*See picture series 184*)

8. DOUBLE WRISTLOCK—HIPLOCK (*See picture series 185*)

9. REAR HIPLOCK (*See picture series 186*)

10. LEG PICKUP—BETWEEN LEGS (*See picture series 187*)

11. LEG PICKUP—TO SIDE—SINGLE (*See picture series 188*)

12. LEG PICKUP—TO SIDE—DOUBLE (*See picture series 189*)

13. SWITCH (*See picture series 132*)
Observe that in illustration 132A you have brought your opponent's arm across your waist, which would be very similar to the position he would place his arms if he secured you with a rear bodylock. As an adversary clasps you in a rear bodylock, reach across your body and secure his arm, hook your leg inside opponent's leg on same side you insert hand in crotch. Twist sharply and the maneuver will then work as described in picture series 132.

All methods of defense and counterattack against an opponent who attacks from the rear must be worked rapidly, as it is difficult to say what opponent intends to do. As you cannot see him, and do not know if he is armed or unarmed, it is best to get out of that predicament in a hurry. Act and ask questions afterward!

In all cases where a bodylock is applied from the rear and your opponent tries to pick you up, a foot inserted and hooked inside your opponent's leg will prevent him from raising you from the deck or throwing you.

Squat—Raise Arms—Turn In—Go Behind—Inside Bar Hammerlock

183A—Opponent has a bodylock, arms are included.

183B—Dropping quickly and raising the arms outward breaks the hold or relaxes it sufficiently for you to continue your tactic.

183C—Turn into opponent, bring near arm over opponent's. Begin to step behind, one arm has already passed behind opponent's back.

183D—Pull out head from bend of opponent's elbow but keep arm in place. Let your arm and hand follow into the spot just vacated by your head.

183E—Place this hand on the shoulder blade. Keeping your body against opponent's forearm prevents him from getting away. Force to deck and attack with free hand if desirable. It will be noticed that this is an inside bar hammerlock (see illustration 110). Applying an upward pressure with your forearm to opponent's, will result in painful injury.

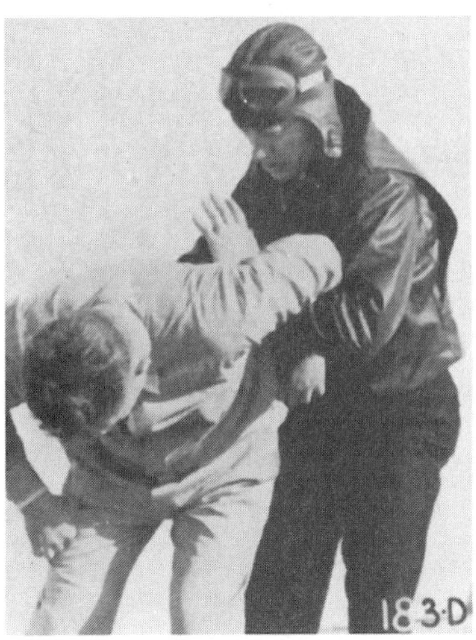

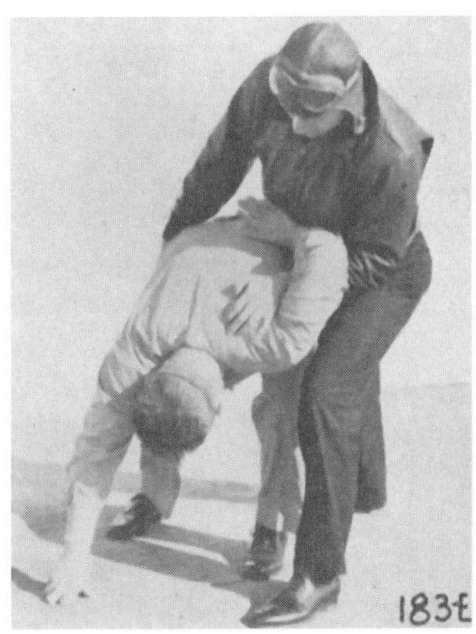

Double Wristlock

184A—Opponent has obtained a bodylock. The arms are free.
184B—Your first action is to reach across your body and secure opponent's wrist with one hand while passing your other arm out and above the elbow.
184C—Inserting the arm over opponent's upper arm and passing it under at the bend of the elbow, it is pushed on through and is locked over your own wrist—thus forming the double wristlock.
184D—With a quick snap outward and back, break opponent's bodylock. At the same time turn out toward opponent's locked arm. The locked arm is then pushed out and back.
184E—Insert your leg in opponent's crotch and continue to push his locked arm up his back; maintain your double wristlock.
184F—Fall backward, at the same time kicking upward into opponent's crotch and force arm up back. Unwary opponent will strike the deck with his face, a continued pressure will break the arm and your foot can deliver additional kicks to the groin. In addition to this maneuver it would be well to study picture series 143. Note the double wristlock is the same, but foot action varies. This wristlock can be used from this position if desired.

Double Wristlock—Hiplock

185A—Opponent has applied a rear bodylock. Obtain the double wristlock as described in illustrations 184B and 184C, preceding.

185B—However, instead of working a complete wristlock shove your hip through to the side the arm is locked.

185C—Snapping down on the arm held securely with the double wristlock will throw opponent over your hip to the deck. Note that opponent his no means of breaking this fall. Drop heavily on his chest as he falls.

Rear Hiplock

186A—Opponent has obtained a bodylock, but has put his head to one side. This indicates that a rear hiplock is in order. Reach across your body, secure opponent's elbow with one hand, quickly twist the body and swing other arm over his head.
186B—Your swinging arm passes over back to opposite arm. It then passes under this arm across chest and locks opponent's arm already held at elbow. (Note position of both hands at this point.)
186C—Quickly stepping forward, use your hip as a fulcrum, heave your opponent over.
186D—Opponent strikes the deck heavily on his back. Shift your arm past opponent's elbow to hold him securely and attack further with kneedrops to the chest, solar plexus or groin.

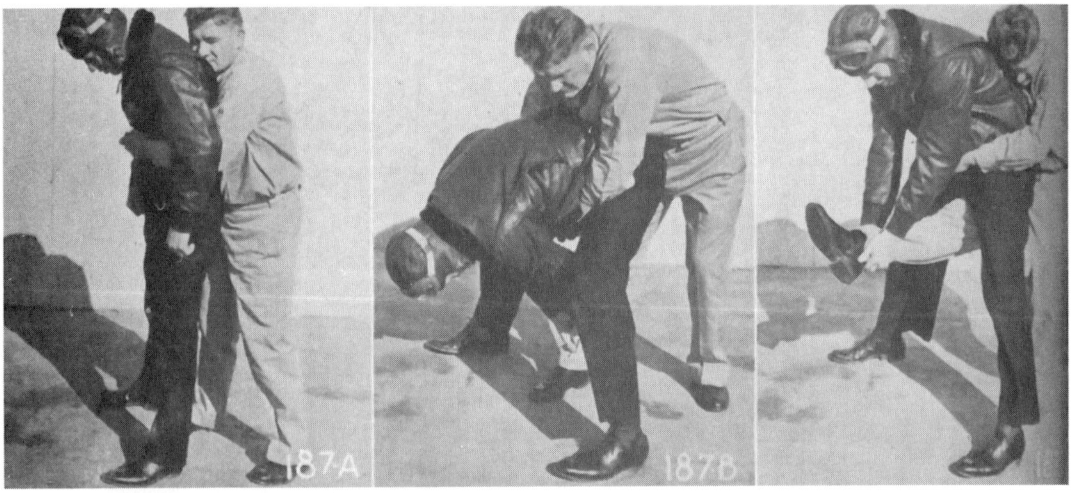

Leg Pickup—Between Legs

187A—Opponent gains rear bodylock and the arms are free.

187B—Spread legs and bend forward quickly grasping one of opponent's ankles with both hands.

187C—Straighten up, retaining grasp on ankle, keeping knee locked and begin to put pressure just above the knee with your buttocks.

187D—Continue the upward pressure sharply and continue the downward pressure on opponent's upper leg; this causes opponent to fall to deck on back. Dropping sharply on his leg just above the knee, will fracture the knee if the grasp is retained on the ankle.

187E—In order to torture victim further shift weight to stomach and snap ankle inward, bending knee slightly, thus an outward pressure is placed against the knee joint. Note illustration 165B; this position can be gained by shifting body.

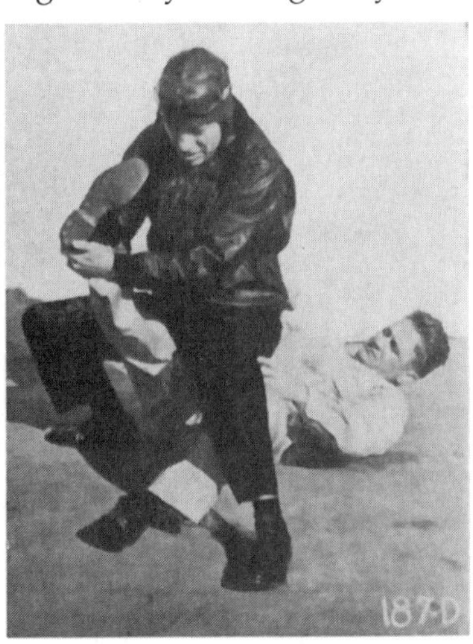

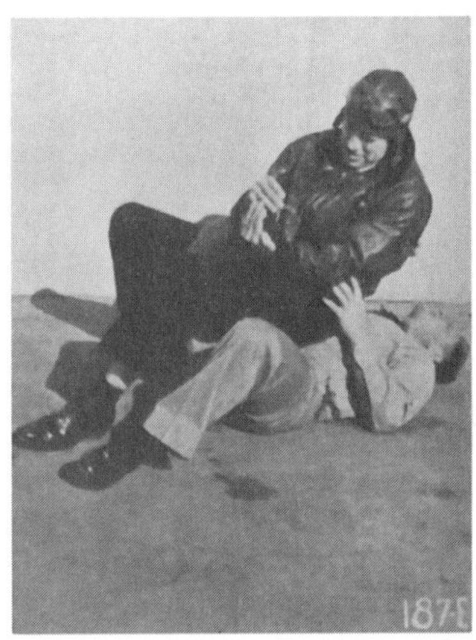

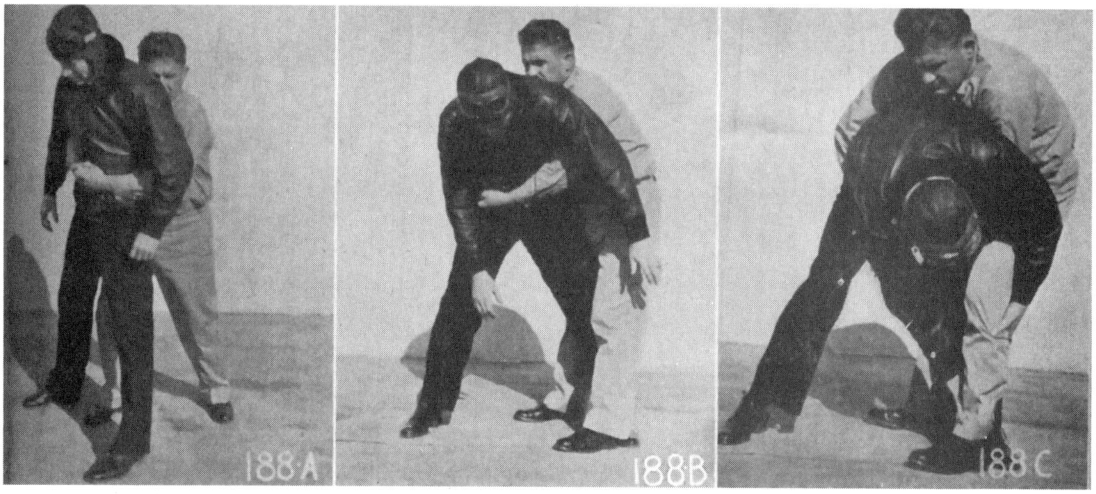

Leg Pickup—To Side—Single

188A—Opponent locks you from the rear under your arms.

188B—Inserting one foot between opponent's legs, by turning slightly reach for his leg.

188C—Grasp the leg, one hand inside ankle, other outside at knee.

188D—Straighten up retaining grasp on leg.

188E—Raise grasped leg and place your leg behind opponent's standing leg and trip backward. Use your knee in groin as he hits the deck.

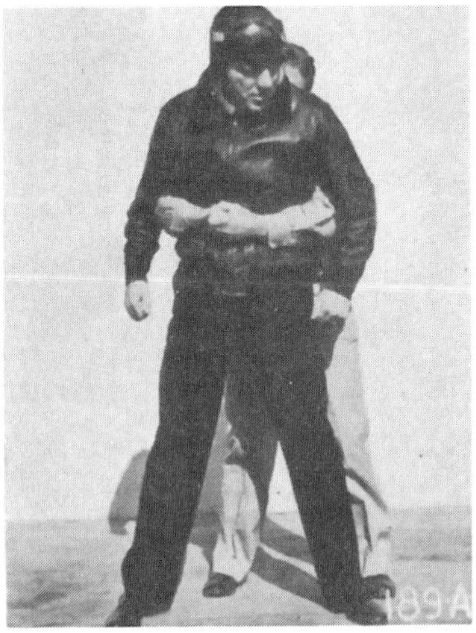

Leg Pickup—To Side—Double

189A—Opponent has you bodylocked from the rear.

189B—Stepping quickly behind his legs and twisting your body at the same instant you are in position to bend forward.

189C—Bend forward and clasp opponent's legs at the knees.

189D—Straighten up lifting opponent from the deck. Now drop him on his back whenever you see fit. Generally, opponent will release his grasp about waist and try to catch himself. If he does not, fall heavily on him as you drop. Note position of hands on opponent's legs.

CHAPTER 9

Searching and Techniques for Control and Leading of Prisoners

THE TASK of searching an armed prisoner when alone is an extremely dangerous one, and should be done with caution. A single slip on the part of the searcher may give a momentary advantage to the prisoner.

Your hand firearm should be held firmly against the hip or at your side, on the side away from the prisoner. Never take your eyes from him for a second; be vigilant and alert at all times. Never attempt a search until the prisoner is at a distinct disadvantage. Never approach within arms-reach until this is accomplished.

If you have assistance in making a search, be sure that you keep out of each other's line of fire. One man should do all the searching while the others stand back and observe closely every movement of the prisoners.

The second part of this chapter is concerned with controlling and leading. It may, at times, be expedient and desirable to control or lead a prisoner without the use of weapons or without inflicting serious injury. These "come-alongs" will enable one to lead a recalcitrant prisoner quickly and easily, with a minimum of effort.

A. Methods of Search

It is well worthwhile to learn the proper techniques of searching and disarming prisoners. There are certain methods which would afford you, as captor, the utmost safety in searching.

1. ON STOMACH, ARMS OVER HEAD (*See picture 190A*)
2. UPRIGHT, ARMS HIGH (*See picture 190B*)
3. KNEELING, ARMS BEHIND BACK (*See picture 191A*)
4. WALL METHOD (*See picture series 192*)
5. PRISONERS FEINTING DEATH OR INJURY
   a. Knee in Neck or Back

If a captured enemy is lying face down, do not turn him over at once. Ascertain, first, if he is dead or wounded. You can protect yourself by placing a knee on his neck or back with your full weight behind it. This prevents the prisoner from rapidly turning over.

b. Bar Toe Hold (*See picture series 193*)
c. Double Bar Toe Hold (*See picture series 194*)

Keep these additional things in mind when searching prisoners:

1. If more than one man have captured prisoners, keep out of each other's line of fire. Let one man do all the searching.
2. Jerk coat collar down to elbows if prisoner is unarmed.
3. Cut belt or suspenders, dropping pants to half-mast.
4. In turning a corner, step well away to side to keep and hold a clear view of all prisoners. Never let one out of sight.
5. If you are not thoroughly convinced that your search has been complete, make prisoners undress, be on the alert for weapons hidden in armpits, crotch and in shoes. Make prisoners who wear breechcloths, take them off; search these thoroughly for small firearms.
6. Look for poisons, hypodermic needles and drugs.
7. Look in the mouth, up nose, in ears, on soles of feet and in armpits.

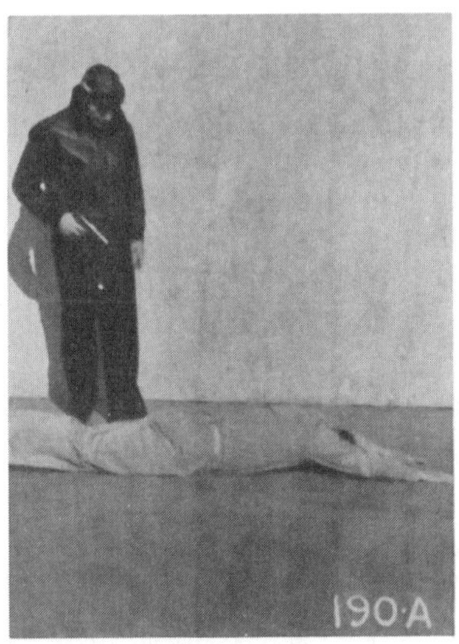

On Stomach, Arms Over Head

190A—Observe that prisoner is face-down with arms stretched well overhead. Prisoners, if more than one, can be searched in this manner. Don't go to each prisoner, make them come to you. Stay at the end of the line. When you have finished searching one man, have him go to end of line and assume same position.

190-B

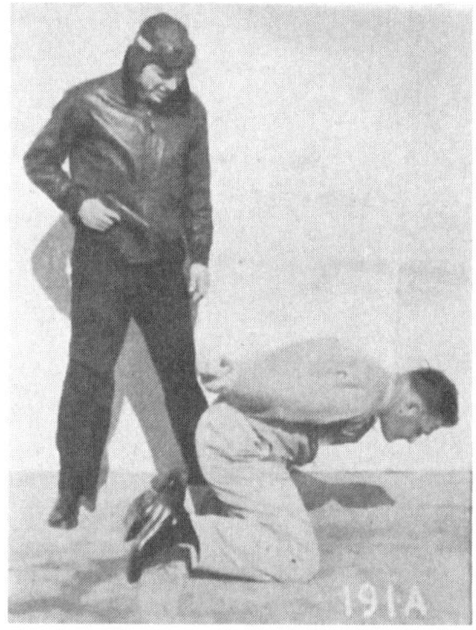

191-A

Ipright, Arms High

'90B—Prisoner's arms are extended fully over head. Approach from rear, slightly to side and keep ,eapon back at side.

:neeling, Arms Behind Back

'91A—Prisoner is at a distinct disadvantage. Approach him from rear. Keep at end of line; if there are ıore prisoners than one, make them change places. Having prisoner place hands on nape of neck is a ariation of the above. Position is precarious and sudden moves are difficult.

Vall Method

'92A—This is the best searching technique (side of tank, automobile or airplane can also be used). With ıe upraised hands against the wall, feet apart and about four feet from wall, all balance is on the hands. .ll prisoners are in a line.

'92B—Holding weapon at side, away from prisoner, place leg in front of prisoner's. Prisoner cannot take ands from wall and if he moves, jerk your leg against prisoner's tripping him to deck. While eyes watch ther prisoners, your hand does the searching. At completion of search, have prisoner move to other end f line and resume same position, you remain at same end. Step back whenever prisoners change places, ut of arms reach.

192-A

192-B

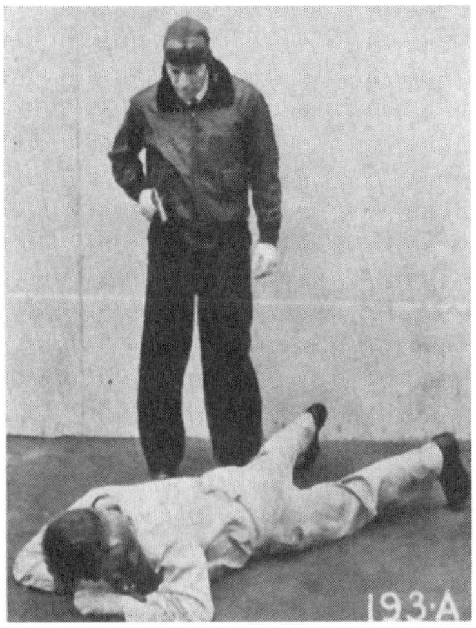

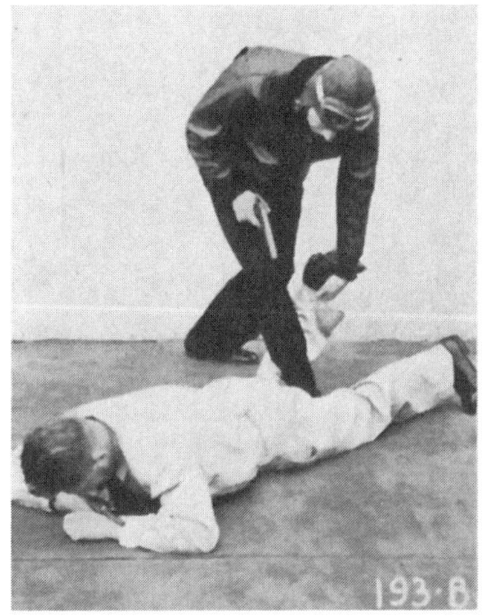

Bar Toe Hold

193A—Enemy is face down, he may be helpless, but never take a chance on that. Approach him from the rear and to side on which he is unable to see your approach.

193B—The action in this, and the following two illustrations is done rapidly. Step over prisoner's leg and pick up ankle with free hand.

193C—Fall forward rapidly locking leg in bar toe hold (note illustration 126) Prisoner's foot is hooked in your crotch and he is thus prevented from turning over to get at you with his pistol.

193D—Twist pistol from grasp, or search from this position if desired. If prisoner persists in any activity, break his leg with forward pressure of your body or strike him over the head with the gun muzzle, never the butt.

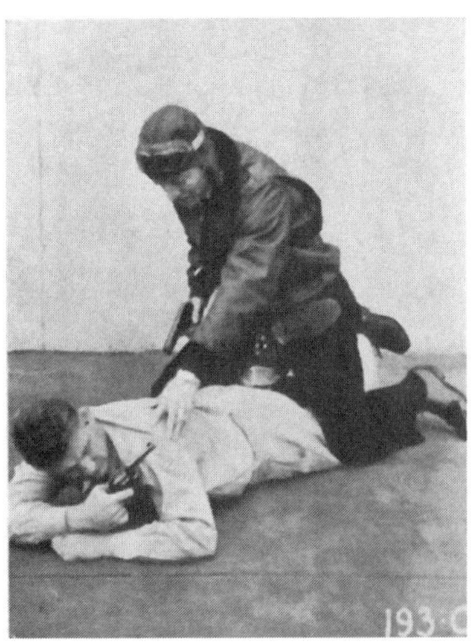

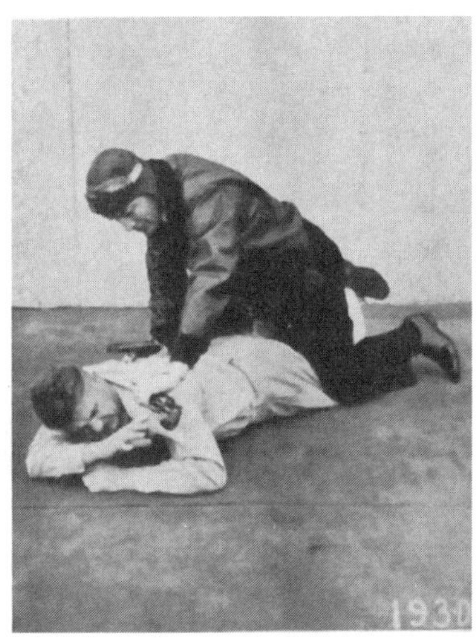

Double Bar Toe Hold

194A—Approach prisoner from the rear and quickly grasp both ankles, place one in the hock of the other knee.

194B—Catching the upraised leg in your crotch, come quickly forward, astraddle your man and instantly reach for his weapon hand if he attempts to use one. Your prisoner may attempt to turn over, but your forward pressure with your body weight will prevent this.

194C—Twist pistol from grasp and strike to his head if necessary. Note how prisoner's legs are locked.

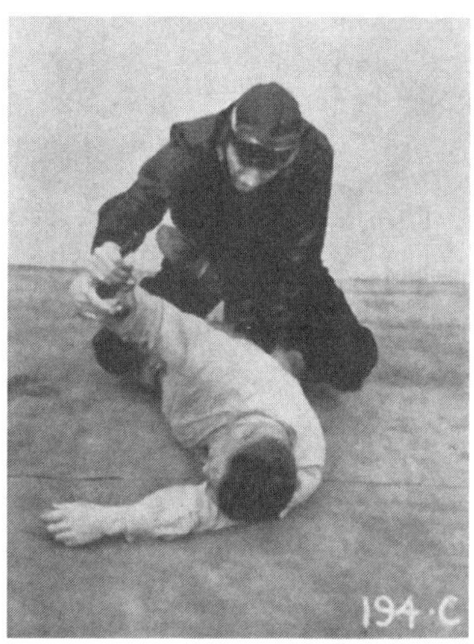

B. Controlling and Leading

(Study maneuvers as listed referring to illustrations as needed.)

1. CROSS WRIST DRAG (*See picture series 195*)

2. CROSS ARM DRAG (*See picture series 196*)

3. HOCK OF KNEE SNAP (*See illustration 78 and picture series 197*)

4. HANDSHAKE (CHEST OR NECK LEAD) (*See picture series 198*)

5. FINGERLOCKS (*See illustrations 93–96 and picture series 199*)

The fingerlocks shown in illustrations 93–96 can be used as leads if you wish.

6. FINGERLOCK—ELBOW TUCKED

    a. Near Arm (*See picture series 200*)
    b. Far Arm (*See picture series 201*)

7. WRISTLOCK—ELBOW TUCKED (*See picture series 202*)

8. WRIST AND MOUTH LEAD (*See illustration 203A*)

9. ELBOWLOCK WITH OPPOSITE ARM WRIST TWIST (*See picture series 204*)

10. OUTWARD ARM TWIST—HAMMERLOCK—WRIST TWIST (*See picture series 205*)

11. CUFF GRASP—HAMMERLOCK

Reach across and grasp opponent's cuff, swing arm out and then snap up his back into a hammerlock. This must be done rapidly as your grip is restricted to a few fingers.

All of the above tactics for controlling and leading a prisoner must be applied with speed. Do not expect your man to wait for you to apply your grip. It is possible that under varying conditions some of these tactics will not work; however, any time a man resists one maneuver, he leaves himself open for another.

Cross Wrist Drag

195A—Grasp opponent's wrist across body (right to left).

195B—Swing arm across front of body and step behind, for control or attack.

Cross Arm Drag

196A—Grasp opponent's arm above the elbow from the inside across body (right to right or left to left).

196B—Swing opponent's arm across front of your body and step behind. If conditions warrant, it is possible to grasp opponent's arm on outside with opposite hand (left to right arm or right to left arm) and snap across front of body, stepping behind for control.

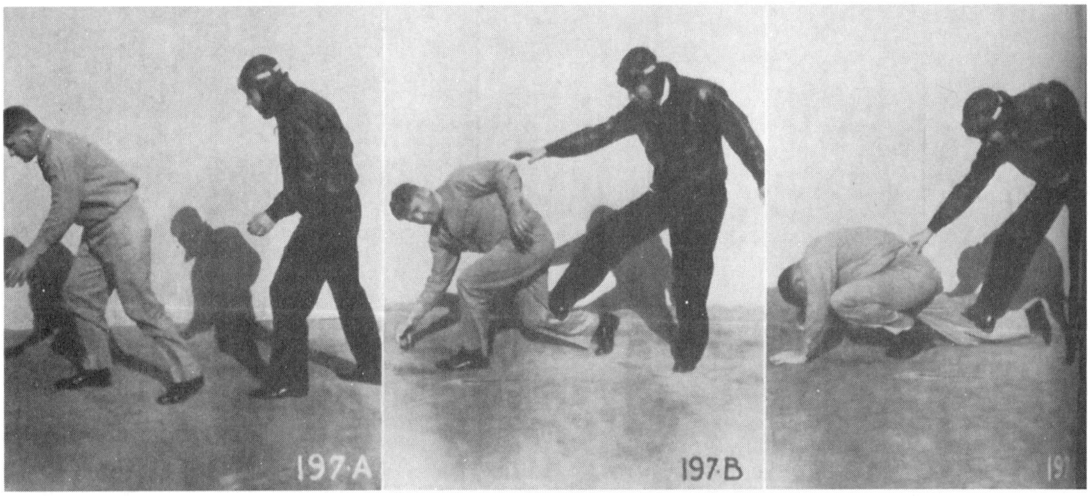

Hock of Knee Snap
197A—Prisoner attempts to escape by turning to run away.
197B—Step immediately into the hock of the knee with your foot, placing weight of body behind it. Thrust hand forward for eventual control.
197C—This forces prisoner to the deck. Your hand has clasped his belt for control and balance. This maneuver closely resembles a football clip. Injury to the knee is often the result of this tactic. This same tactic may be used by stepping into the hock of the knee of a fleeing man if pursuer wishes to bring his man down. As the running fugitive is approached, jump into back of leg.

Handshake (Chest or Neck Lead)
198A—From a handshake position as shown, or by grasping the inside of opponent's wrist you can easily get a control lead.
198B—Retain grip on hand or wrist and pass other arm under held arm and over chest to opposite side. Grasp clothing for security and press down on arm, keeping opponent's wrist uppermost, locking elbow. Lead him off!
198C—Retain same grip on hand or wrist, pass your other hand under held arm and behind the neck. Pull held arm down over lever arm; this not only locks the elbow, but exerts a pressure on opponent's neck. He is now in position to be led off.

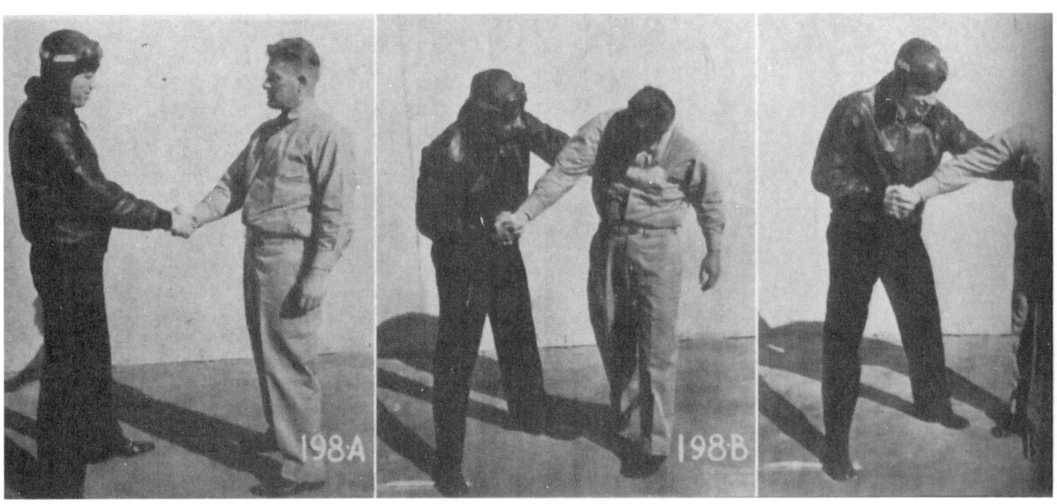

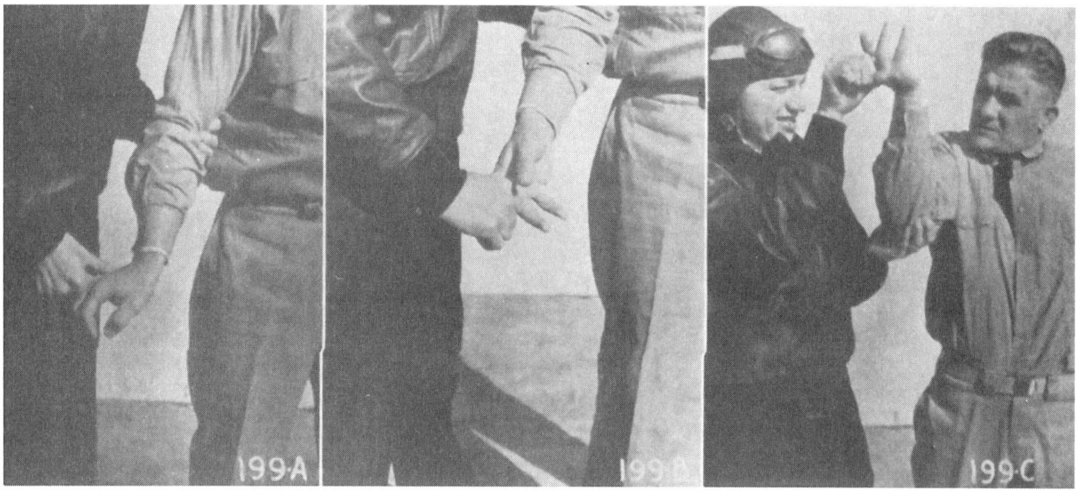

ingerlocks

99A—Grasp opponent's elbow with one hand, being sure to keep it locked straight. Insert your other and between the ring and middle fingers of opponent's hand, grasping ring and little fingers. Place humb on back of opponent's hand and press down as you twist outward. Study the illustration for erfection of this maneuver.

99B—Place one hand on opponent's elbow and grasp his fore and middle fingers from the inside with our palm down. (Note illustration)

99C—Quickly raise forearm and press down on fingers. Hand under opponent's elbow prevents him om lowering it. Lead off!

ingerlock—Elbow Tucked—Near Arm

00A—Grasp ring finger or any finger with a fingerlock from behind palm as shown in illustration.

00B—Raise opponent's arm and tuck it in the bend of your elbow with your free hand.

00C—Apply pressure down on finger back toward elbow for an effective "come along." Do not worry bout his other hand as he would not dare use it since you can easily break the held finger.

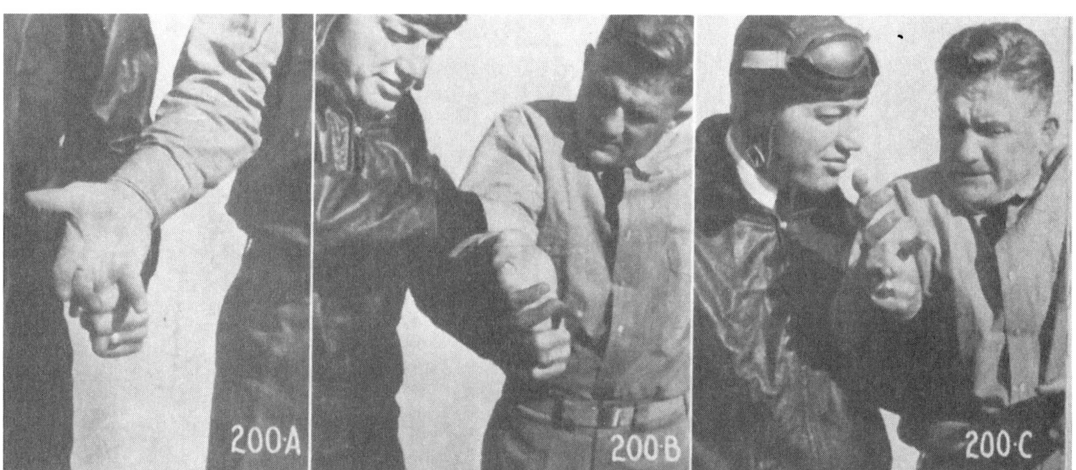

Fingerlock—Elbow Tucked—Far Arm

201A—Secure opponent's forefinger as shown in illustration or any finger with a fingerlock.

201B—Raise held finger and insert arm above elbow, tuck in securely.

201C—Now apply pressure down and back on held finger, use other hand at the wrist for greater security. The other hand could grasp opponent's little finger and apply a finger spread from this position. (See illustration 95)

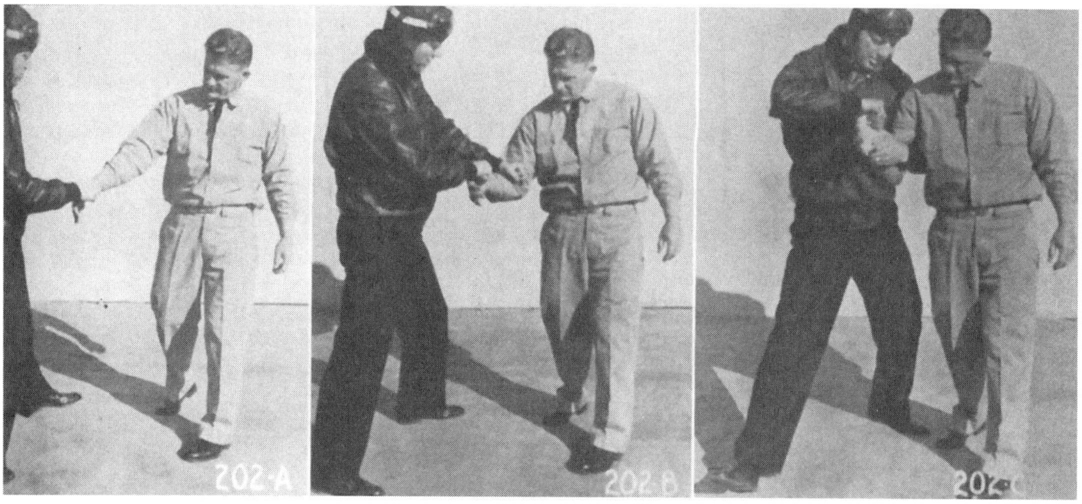

Vristlock—Elbow Tucked

02A—Grasp opponent's band over thumb, your thumb to back of his hand.

02B—Strike sharply to inside of elbow with edge of fist or hand to bend it, if resistance on opponent's art is apparent.

02C—Stepping to side of opponent, pass your free arm under his held arm just above the elbow and ecure his wrist from the inside.

02D—Snap opponent's wrist in and down sharply.

02E—Adjust both hands over opponent's wrist as shown and press down on elevated wrist. Do not ress straight back as opponent may get elbow free. Elevating the wrist higher than his elbow prevents iis. This is an excellent "come-along." Pain to wrist is severe.

Vrist and Mouth Lead

03A—Stepping behind opponent and shifting his tucked elbow to other side of your body from lustration 202E above and using your free hand to insert a finger inside opponent's cheek will give you iis effective lead. Apply pressure to the finger inserted in mouth turning opponent's head as far as ossible; this alleviates any chance of his biting your finger.

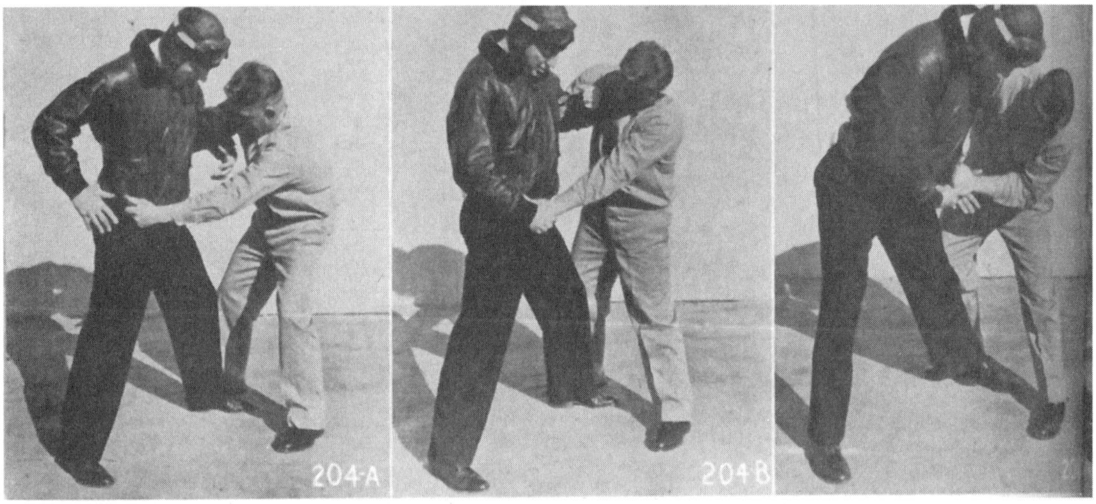

Elbowlock with Opposite Arm Wrist Twist

204A—If an opponent attempts to bodylock you from the front, you can apply this excellent lead c throw.
204B—Elbowlock opponent's far arm and clasp his near arm at the wrist, your thumb passing across th inside of his wrist.
204C—Bring your far arm across between your body and opponent's body. Force opponent's held wris down and back within range of this arm and hand. Grasp it from the underside.
204D—Your fingers pass under the wrist and come up across palm from thumb side. Pressure i downward. Your thumb is against inside of wrist.
204E—Your free hand can grasp opponent's arm which passes across your back if desired. Notice tha both of opponent's arms are held with one arm. Additional pressure on wrist and hand will caus opponent to bend over where he is in position for a kneelift to the face.
204F—Opponent can be thrown with a hiplock from this position if desired. Just release grasp on wris and clasp elbow with free near hand, shove hip through and throw to deck. (See hiplock—regula illustration 120.)

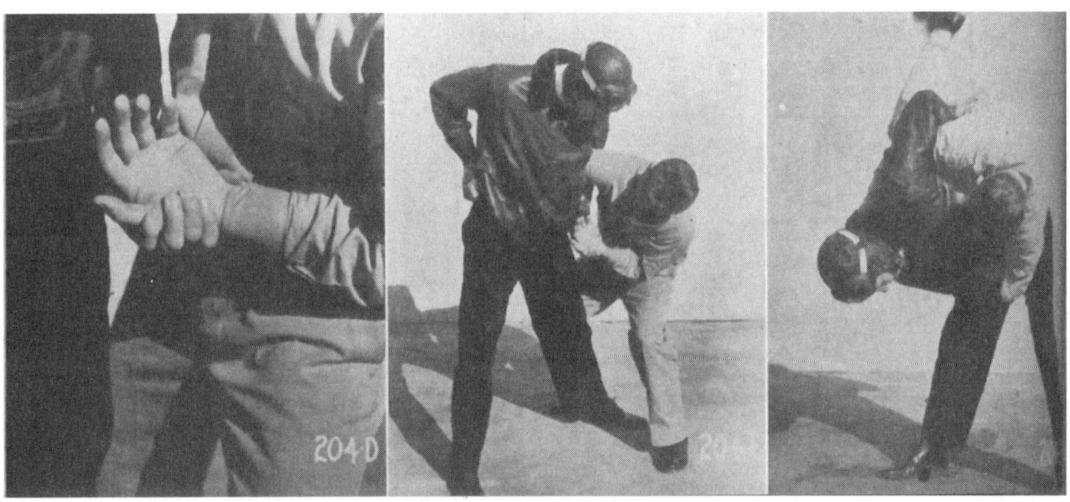

Outward Arm Twist—Hammerlock—Wrist Twist

205A—Clasp opponent's wrist as shown, the near hand at wrist, other hand above.

205B—Step in and swing opponent's arm up with a jerk and begin an outward turn.

205C—Passing opponent's upraised arm over the head continue your turn.

205D—This brings you behind your man. Maintain grip on wrist and place other hand on elbow.

205E—Now snap arm up opponent's back to a regular hammerlock and turn wrist down to a twist (see illustration 109). Apply pressure against the wrist and elbow for the lead. Push opponent forward.

CHAPTER 10

Disarming

THE TACTICS and maneuvers listed in this chapter are without a doubt very important for the fighting man, and concern not only disarming techniques of firearms, but the club and knife as well. The disarming combinations described herein are devised for emergency use, when quick, sudden action may be necessary to save a life. These combinations are not intended to be used under less severe circumstances, such as saving a few dollars from a "stick-up" man. In other words, these disarming maneuvers are devised to give the fighting man a chance for his life when he knows that death or imprisonment will result if he submits or resigns himself to his fate.

Many disarming combinations have been developed against all types of weapons, and probably a complete manual could be prepared on this subject alone. No attempt has been made here to include all the maneuvers and "tricks" devised, but to present in a clear and understandable manner to the fighting man those movements which are practicable and quick in their application.

It will be noted that in many of the combinations, the first movement necessitates very quick action, such as getting out of the line of fire or striking the weapon to one side. You have an advantage here because you know what you are going to do, but your armed adversary has to react to the situation. Therefore reaction time plays an important part in pistol and rifle disarming. The man with the firearm is sure of himself, but for him to squeeze the trigger on his weapon, he must first see what you are doing. Although reaction time is short, it is not as short as the time it takes you to get out of the line of fire or strike the weapon away.

The latter part of this chapter deals with disarming tactics of enemies with knives and clubs. Again, it will be noted, that you do not retreat, but close in on your adversary, thus cutting down his range of action.

Peculiarities of small firearms are well worth knowing, especially your own.

The .38 and .45 Colt Automatics will not fire when pressure is exerted against the muzzle of the weapon. This is caused by a slight displacement of the slide and barrel of the gun. However, smaller automatics being built on a different scale are not affected by this pressure. Yet, if the slide is firmly or pushed back, the weapon cannot be fired, nor can the weapon be fired after the first shot because it cannot reload.

The German Luger Automatic Pistol, generally carried by officers, though greatly different in appearance from our automatic pistols, reacts in the same manner as do the .38 or .45 Colts. However, the reloading spring in the Luger is approximately three times as heavy and, therefore, a correspondingly heavier pressure must be exerted against the muzzle.

The Japanese have an automatic pistol very similar in make to the Luger and it is believed to be a copy. It is presumed that this pistol would react to a pressure on the muzzle as does the Luger; however, the authors not having experimented with this weapon cannot make a definite statement to this effect.

Any revolver, with a visible hammer, when not cocked, can be kept from firing by quickly grasping the cylinder firmly with one or both hands. Do not try this on hammerless revolvers.

When confronted by an assailant with a peculiar type firearm, it is best to use the standard disarming tactics rather than take a chance with the unfamiliar weapon.

The above peculiarities are mentioned to give additional hints as to what to do in cases of dire emergency or where you cannot disarm an enemy.

206A—This is a .38 or .45 Colt Automatic, fully cocked with safety off.

206B—Notice that when pressure is exerted against the muzzle the slide moves back a short distance. This prevents the gun from being discharged. The grip on the wrist makes your hold firm. Notice position of the notches on side of slide in this and previous picture 206A. Try this on your own automatic. Be sure your weapon is unloaded at the time.

206C—Grasping the revolving cartridge chamber of a revolver prevents gun from being fired when weapon is not cocked. In order to fire a revolver the cylinder must turn one notch before hammer falls. Do not attempt this on a hammerless revolver.

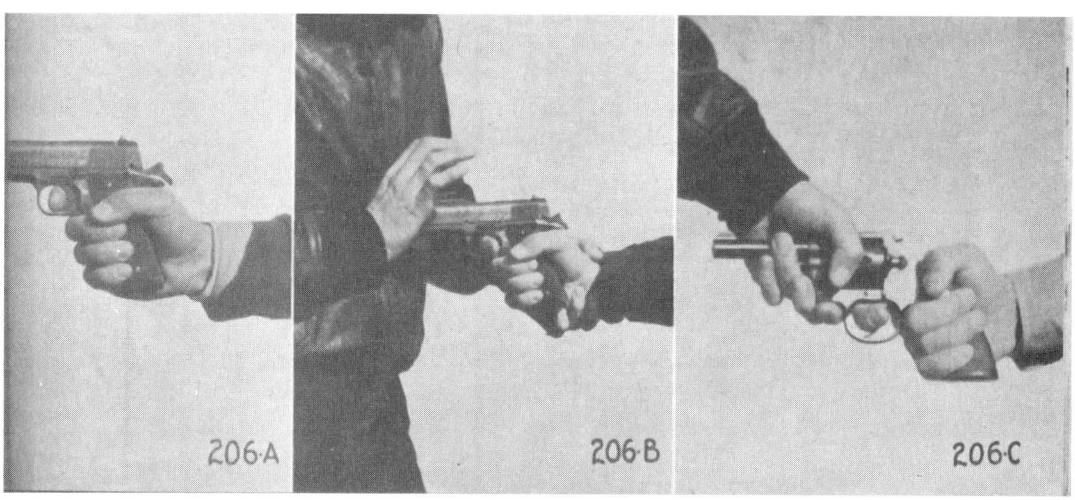

A. Pistol

I. Shoulder or Cross Draw Holster

1. REVERSE WRISTLOCK (*See picture series 207*)

2. FACE STRIKE (*See picture series 208*)

II. Side Holster

1. OUTSIDE BAR HAMMERLOCK (*See picture series 209*)

2. DOUBLE WRISTLOCK

Reach straight across, secure your opponent's drawing wrist, and apply a double wristlock. The rest of the maneuver may then be worked as illustrated in picture series 143.

III. Pistol Drawn (Hands Raised)

1. INSIDE TURN (*See picture series 210*)

2. OUTSIDE TURN (*See picture series 211*)

3. SCISSOR SWING (GUN IN FACE) (*See picture series 212*)

This entire action takes only a fraction of a second to accomplish. The maneuver is worked so fast, that even though the pistol may be cocked, opponent loses the weapon before he can pull the trigger. A person's reaction time, at best, is never lower than 1/4 of a second and is usually 3/8 of a second or more. The entire action in this tactic is less than 1/4 of a second.

4. SCISSOR SWING (GUN IN STOMACH) (*See picture series 213*)

5. SHOULDER GRASP (EMERGENCY ONLY) (*See picture series 214*)

IV. Pistol Drawn (Hands Down)

1. UP AND DOWN SCISSORS (*See picture series 215*)

2. INSIDE TURN (*See picture series 210 and 216*)

The tactic in picture series 216 is accomplished in the same manner as that shown in picture series 210, except that the starting position is different.

3. OUTSIDE TURN (*See picture series 211 and 217*)
Execute the tactic shown in picture series 217 as that shown in picture series 211. The only variation is at the start of the maneuver.

4. ARM STRIKE—GUN AT TEMPLE (*See picture series 218*)

V. Pistol in Back

1. OUTSIDE TURN (*See picture series 219*)

2. INSIDE TURN (*See picture series 220*)

Reverse Wristlock

207A—As opponent attempts to draw a pistol from a shoulder holster, strike his elbow with an upwar and outward thrust of the far hand.
207B—An opponent reaching for a shoulder weapon puts himself in an awkward position, so afte striking his elbow insert your other hand up between his arm and body and secure his wrist from th inside.
207C—This may not prevent him from drawing his weapon, but this does not make any difference. Brin the hand which struck opponent's elbow forward to grasp the wrist also.
207D—With a quick snap, jerk the weapon arm downward and back, maintaining your grip on the arm.
207E—Continue the motion back, which locks elbow and wrist. It will be noted that the pistol at no tim is pointed toward you as the disarmer.
207F—Securing opponent's wrist with your far arm, take gun from grasp and strike to back of head wit muzzle, or kick to face or body. This maneuver works exactly the same if opponent draws from holste across waist.

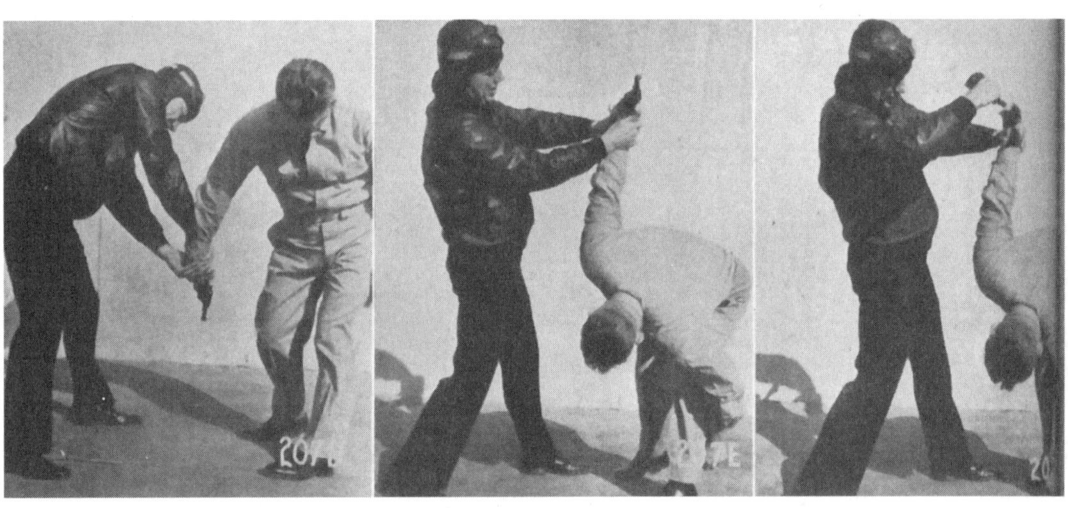

Face Strike

208A—Opponent starts draw from shoulder.

208B—The elbow is struck exactly as in illustration 207A, opponent's arm thus being forced outward.

208C—Maintaining grip on opponent's elbow, use your other hand to grasp his wrist and snap pistol sharply into his face.

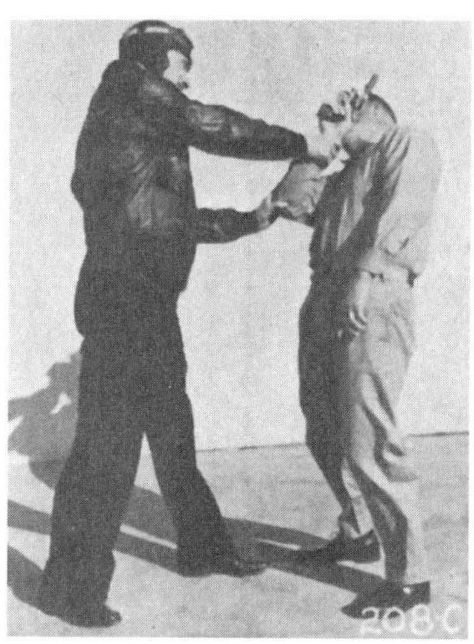

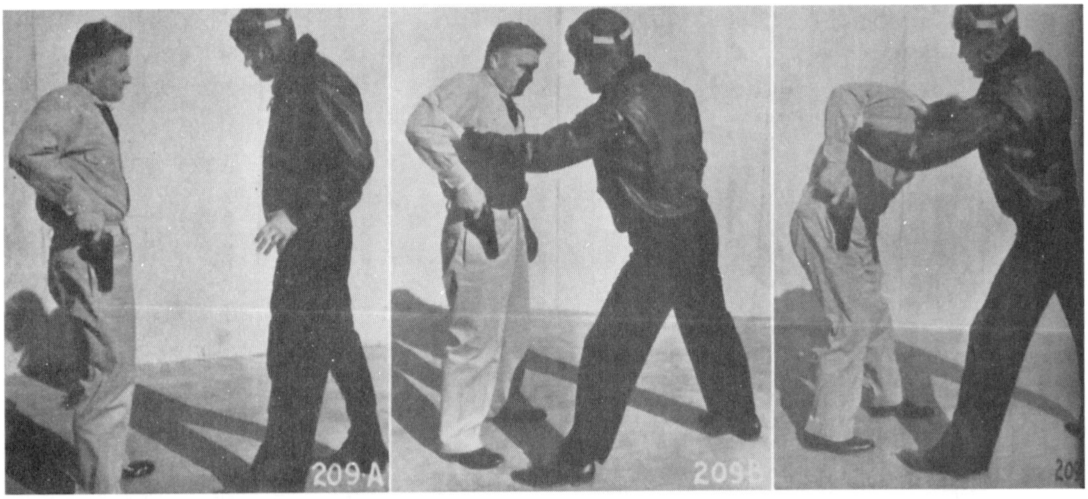

Outside Bar Hammerlock

209A—Opponent starts a draw from a hip holster (as shown) or from hip pocket. Notice the bend in his arm at the elbow.

209B—Quickly insert your hand, palm up, into the bend of opponent's elbow as you step forward. Thumb is outboard.

209C—For additional leverage place your other hand on his shoulder while pushing into bend of elbow.

209D—Now turn completely around, holding your grip in the bend of opponent's elbow. This brings opponent's arm into an outside bar hammerlock. Your upper arm and body are kept close to opponent to prevent escape; this forces him to the deck.

209E—Reach up and pull pistol from his grasp and strike to head if you wish. Continued pressure to the barred arm will break it. If you completed this maneuver before your opponent got his hand on his weapon, draw it yourself from this position with your free hand.

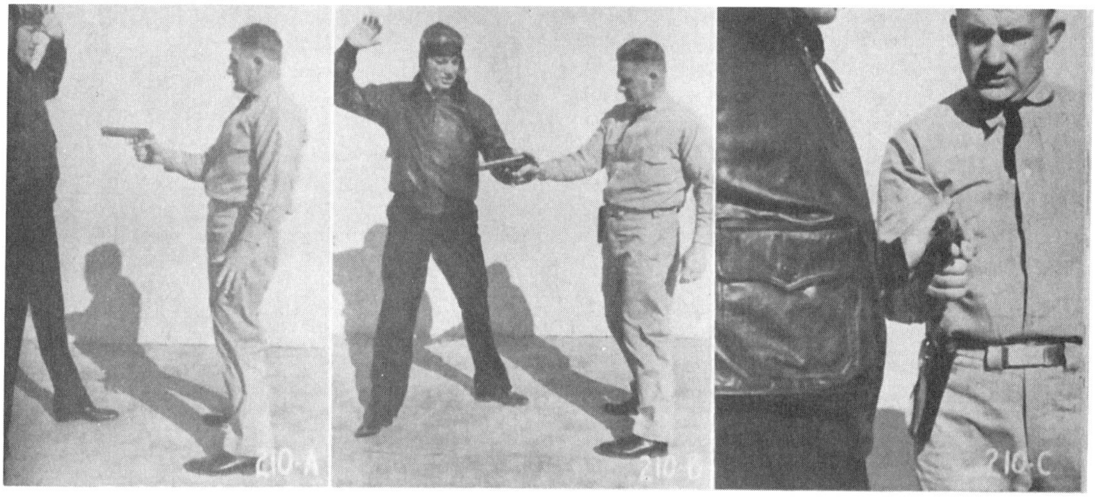

Inside Turn

210A—Opponent has pistol drawn, your hands are raised.

210B—Twisting the body quickly to the inside, strike the gun hand away with the far hand. This is the most important move of this maneuver and must be executed with speed.

210C—This illustration shows that you are now out of the line of fire. It is a side view of illustration 210B above. Reaction time of any person allows you to strike and twist from line of fire before your opponent can pull the trigger.

210D—Now quickly bring down your other hand against the side of the pistol and force inward toward opponent's body.

210E—Continue this action, any discharge of the pistol now will only injure your assailant.

210F—By pulling on opponent's wrist and pushing on the pistol, weapon is forced from his grasp. Step back, retaining grip on wrist for control if necessary at the moment.

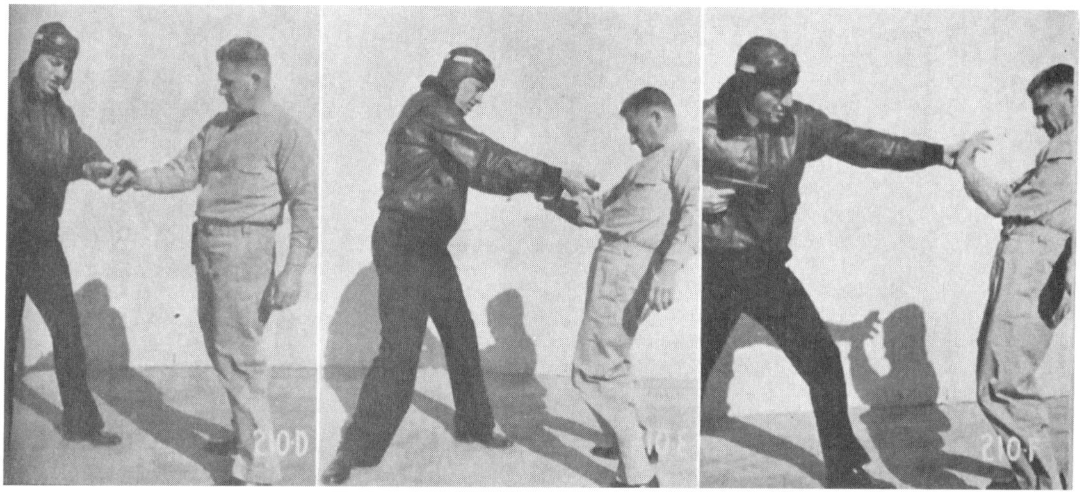

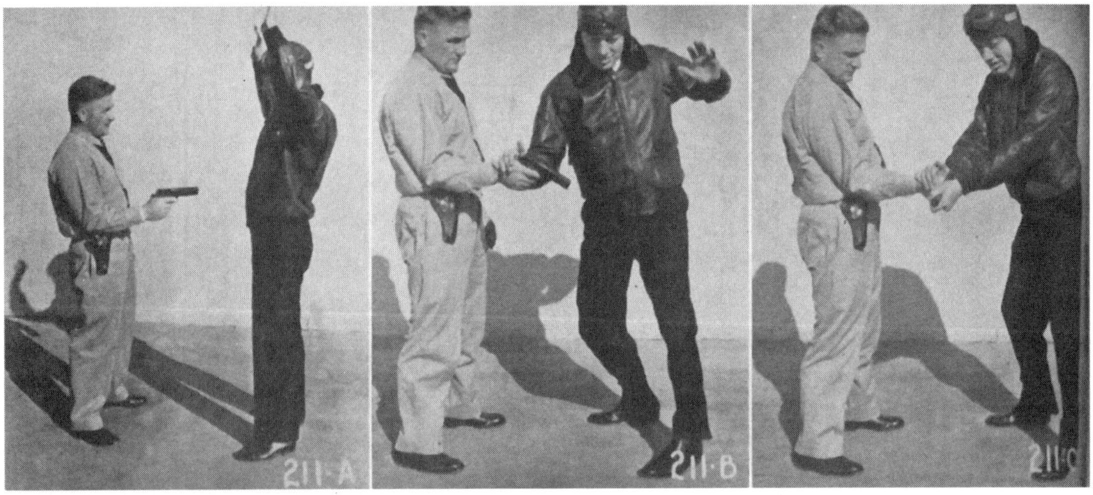

Outside Turn

211A—Opponent has pistol drawn, your hands are raised above head.

211B—Quickly turning the body outward, strike the gun hand at palm with a downward sweep of the arm. This maneuver is most essential, as you must get from the line of fire as was illustrated in 210B and 210C above. Because of the fact that you act quickly and your opponent must react, he cannot pull the trigger in time to put a bullet in your body.

211C—Immediately after striking the pistol, bring the other hand quickly down and grasp the pistol and twist outward toward trigger finger and back of opponent's hand.

211D—Continue twist, locking opponent's trigger finger in trigger guard. Added pressure will bring him to the deck where he is in position for kicks to the face, chest, solar plexus or groin.

211E—A close-up of illustration 211D. Note how opponent's trigger finger is locked in guard. Observe position of your hand on the barrel of the automatic. This same maneuver is effective with a revolver. Finger can be easily broken or practically torn off by jerking down on the weapon when in this position.

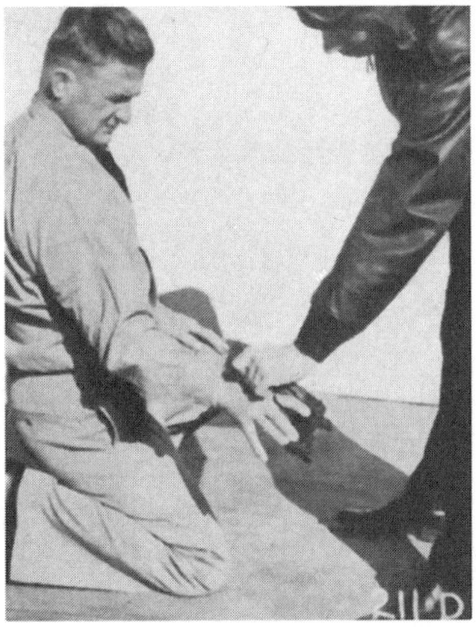

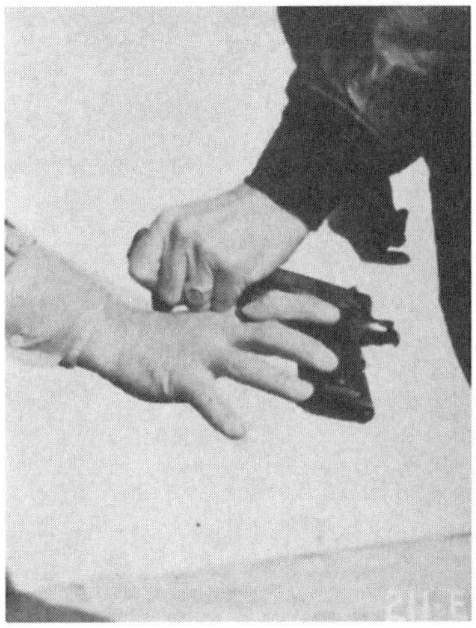

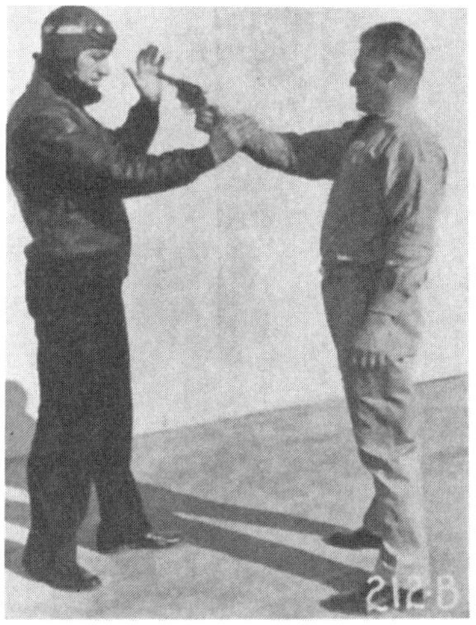

Scissor Swing (Gun in Face)

212A—Opponent puts pistol in your face; your hands are raised overhead.

212B—Notice this action: Quickly bringing your hands down, one strikes the inside of the wrist while the other strikes pistol from opposite side. This scissor action forces pistol inward against opponent's fingers.

212C—Continuing the scissor action the pistol begins to turn in opponent's hand.

212D—Carrying through with swing knocks pistol from grasp. Do not try to pick up weapon. Let opponent go for it and when he does, kick him or deliver a kneelift to his face. (Observe this maneuver in Hand-to-Hand Combat Training Film.)

Scissor Swing (Gun in Stomach)

213A—Opponent has pistol in stomach; your hands are raised.

213B—Scissor swing is accomplished in the same manner as the swing illustrated above in picture series 212. Hands are brought rapidly down and moving in opposite directions.

213C—One hand strikes inside of wrist, while other hand strikes pistol.

213D—This scissor swing strikes pistol from opponent's hand by twisting it against his fingers. (Note that the strike is never done in the opposite manner, a strike to the back of the wrist locks gun in hand.) Entire maneuver takes less than 1/4 of a second to work.

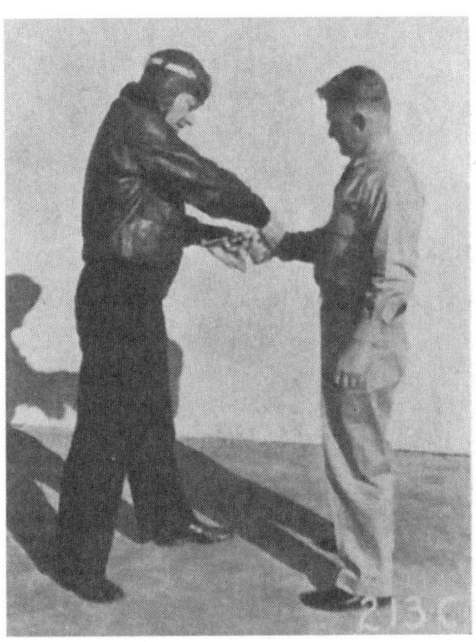

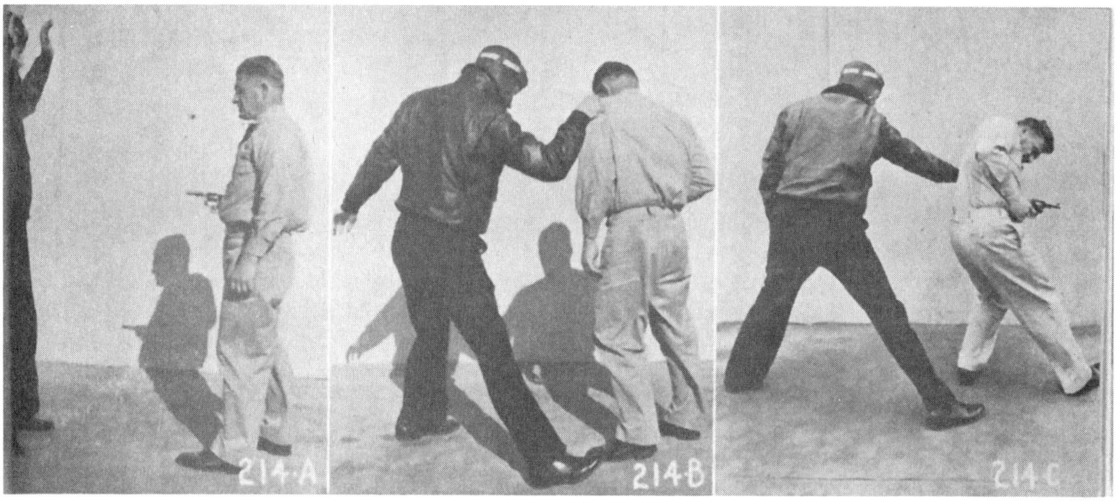

Shoulder Grasp

214A—Here is a situation that places your opponent at a distinct advantage. He has the "drop" on you and is smart enough to keep his weapon at his side. Yet if you know that he is going to kill you, you just cannot resign yourself to your fate, so let's do something; let's take a chance.
214B—Step forward quickly and grasp opponent's clothing at the shoulder and spin him away from you, thus forcing his pistol out of line of fire of your body.
214C—Swing him as far as possible away from you. Note position of his gun arm.
214D—Reach through the bend of opponent's elbow and secure his wrist from the inside with the near hand and place the far hand over his wrist on the outside.
214E—Snap his arm up to the rear sharply. Actually this maneuver develops into a reverse wristlock.
214F—Securing opponent's wrist with the far hand, twist pistol from his grasp, shoot him or strike him with it. From the moment you swing out of the original line of fire, the pistol is never again pointed at you. This entire action takes less than a second to accomplish. The danger lies in the first maneuver; it must be done with utmost speed.

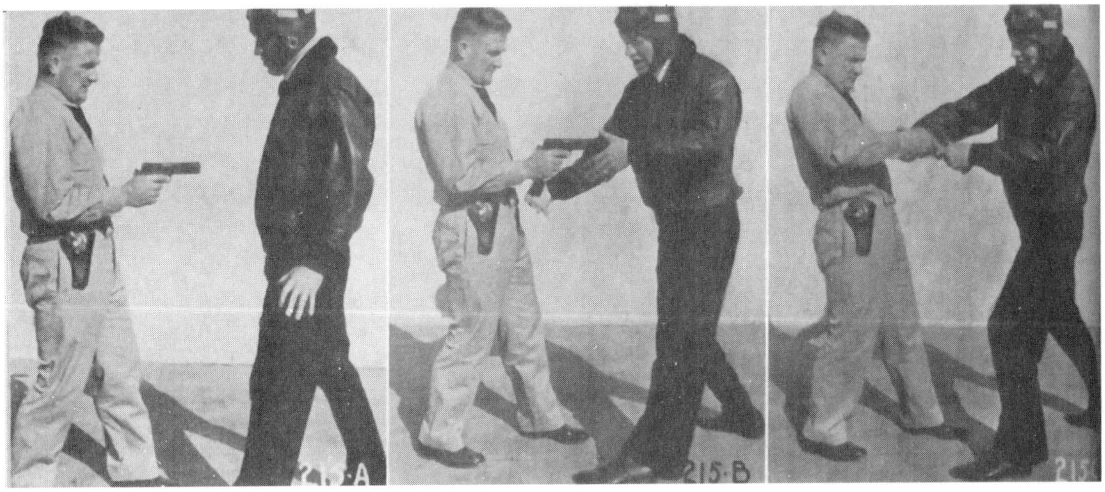

Up and Down Scissors

215A—Opponent draws gun and orders you to raise hands. You are inclined to comply, or lead him to think that you are.

215B—Start to raise hands, but bring them quickly forward, at the same time twisting away from line of fire.

215C—Your far hand comes up under opponent's wrist either with a grasping or striking motion. Your near hand comes up just high enough to grasp the barrel of the pistol. Action is up on the wrist and down and out on the pistol.

215D—A continuance of this scissor action sweeps pistol from opponent's grasp. It is impossible for him to retain his grip due to the pressure on the wrist upward.

215E—Snap pistol from grasp and place in own hand ready for use.

215F—Step back, pistol at side ready for use, other hand on alert to ward off any attack. This entire maneuver is worked with great rapidity; it takes less than 1/4 of a second to accomplish. However, practice is essential to become proficient in its operation.

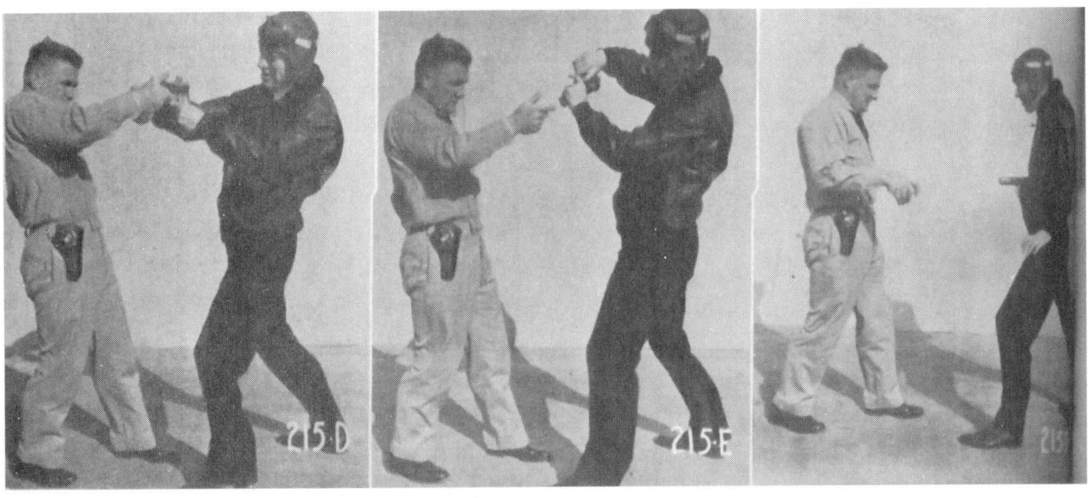

Inside Turn

216A—Opponent draws pistol and orders you to raise your hands.

216B—But instead turn in and strike back of opponent's pistol hand with your hand, thumb up. From this position carry on as in illustrations 210D, 210E, and 210F.

Outside Turn

217A—Opponent draws pistol and orders you to raise hands.

217B—Turn quickly outward, out of line of fire, striking opponent's pistol hand sharply on inside of wrist. Your hand has thumb up. From this position continue to disarm as shown in illustrations 211C, 211D, and 211E.

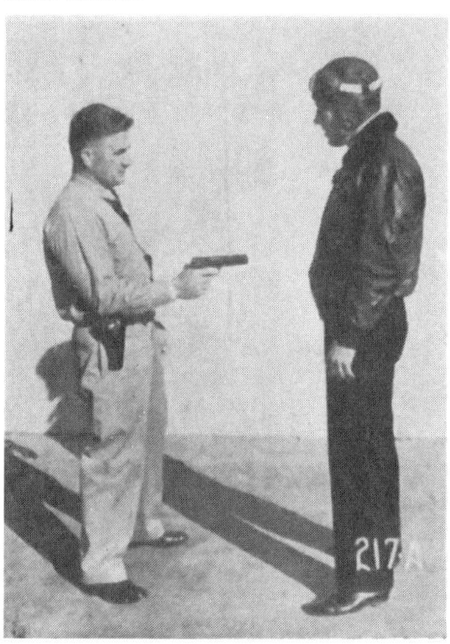

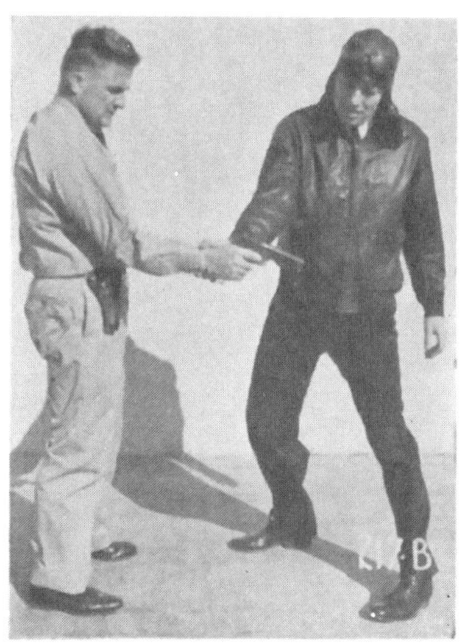

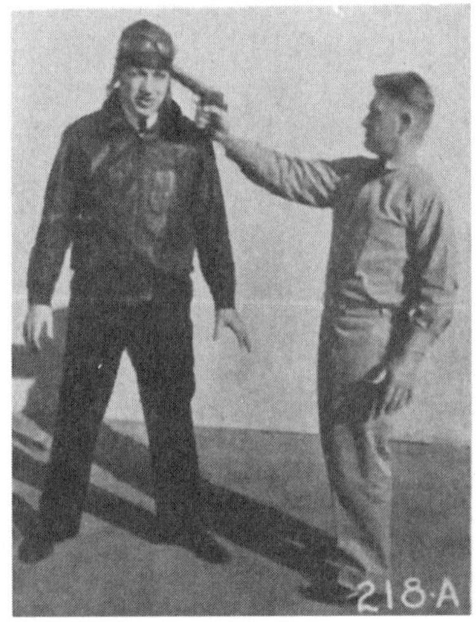

Outside Turn (additional)

217C—From the position, illustrated in picture 217B, bring hand up and lock trigger finger with you thumb. Fingers pass over back of opponent's hand. This pressure pushes opponent's trigger finge against front of trigger guard and prevents pistol from being discharged. Retaining this lock, attac opponent with kicks to his groin. (Type of pistol does not vary maneuver.)

Arm Strike—Gun at Temple

218A—Opponent places gun at temple.

218B—Throw arm up quickly, sharply striking opponent's arm at the wrist, turning body at same time This gets you out of line of fire. In practically all cases pistol will be knocked from opponent's grasp.

218C—Carrying through with your arm strike, kick hard to opponent's groin.

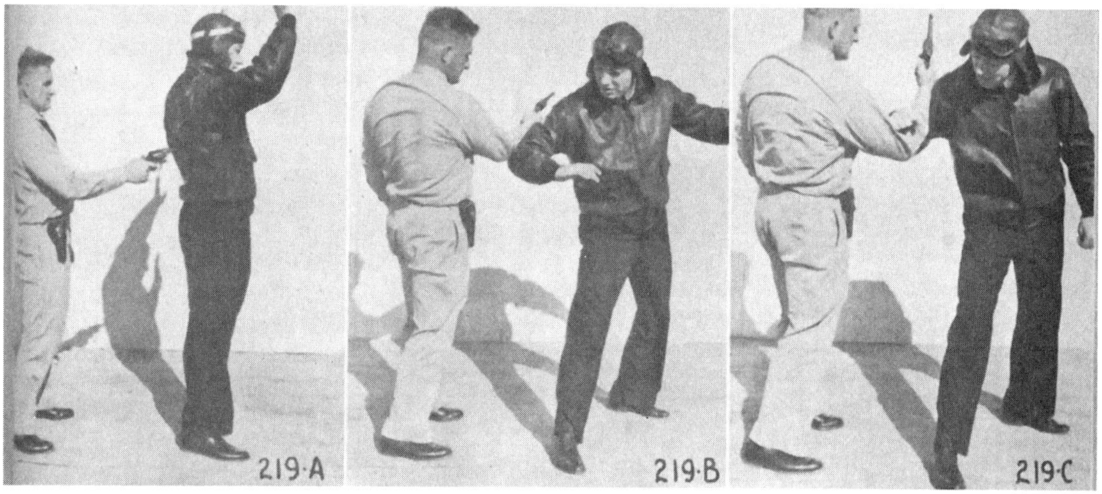

Outside Turn

219A—Opponent places pistol in your back. Turn head, note which hand holds pistol, and then raise arms in surrender.

219B—Turn outside quickly, stepping back slightly and at the same instant strike opponent's pistol arm with your elbow, sharply.

219C—Pass the arm which struck opponent's pistol arm under and bring hand over to lock on inside of opponent's elbow. This raises opponent's forearm.

219D—Use your shoulder to press against the forearm and other hand for additional leverage. Force opponent's hand up to his shoulder.

219E—Snap weapon from opponent's grasp by pulling down on barrel with free hand. Note that muzzle of pistol never points at you after first turn which gets you quickly from line of fire.

Inside Turn

220A—Opponent places pistol in back. Again turn head, note hand holding pistol, and raise arms in surrender.

220B—Turn inside opponent's pistol arm suddenly, passing your arm over it. Note that this quick turn immediately gets you out of line of fire.

220C—Secure opponent's weapon arm just above elbow and lock this arm to your side. Step into your man—

220D—And deliver a heel of hand blow to the jaw (or any other facial blow), and a kneelift to the groin. Don't worry about the pistol, its useless in this position.

B. Rifle

(Bayoneted or Unbayoneted)

I. Facing Opponent

1. RIFLE TWIST (*See picture series 221*)

2. RIFLE STRIKE AND GRASP

The situation is the same as shown in illustration 221A. Strike rifle outward and grasp. Secure it firmly and kick to opponent's groin, shins or knees.

3. SNAP BAYONETED RIFLE TO DECK (*See picture series 222*)

4. RIFLE SNAP

The rifle is struck to the side as shown in illustration 221A and the muzzle is seized with one hand while the other hand grasps the butt or breech as shown in illustration 221B. As the rifle is twisted across front of opponent's body as shown in illustration 221C, your far leg comes across in front of your opponent. Thus you wrest the rifle from your opponent's grasp by using your hip as a lever. This maneuver is practicable when opponent is not inclined to resist or pull back and does not necessitate the additional movements as shown in illustrations 221D, 221E and 221F. Instead of snapping butt into face, snap muzzle.

II. In Back

1. OUTSIDE TURN (RIFLE HIGH IN BACK) (*See picture series 223*)

2. OUTSIDE TURN—HIPLOCK (*See picture series 224*)

3. OUTSIDE TURN—RIFLE TWIST

Opponent has rifle in your back, turn outward, step back, strike rifle barrel with elbow. Grasp rifle barrel from underside and breech from top. Complete maneuver as outlined and described in picture series 221. Your turn places you in position to grasp rifle as shown in illustration 221B.

4. INSIDE TURN (*See picture series 225*)

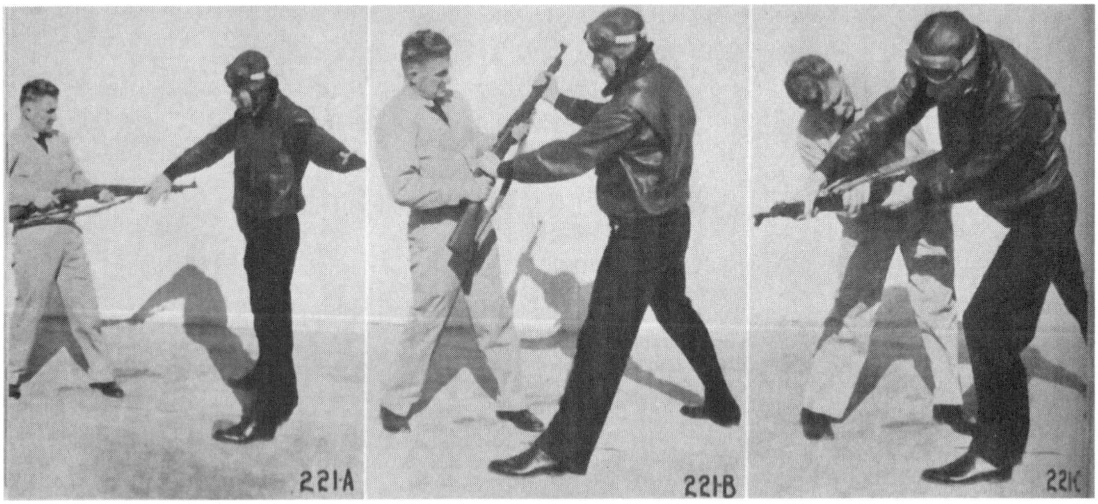

Rifle Twist

221A—Opponent has a rifle on you, your hands are raised. Bring down one hand quickly to the inside of the barrel, palm striking rifle outward and step in. Other hand follows through and—
221B—Grasps rifle at breech. Striking hand grasps barrel near muzzle. Note that one hand passes over breech while other hand clasps barrel from underside. This hand position is identical to opponent's clasp.
221C—Twist rifle across front of opponent's body. His inclination will be to resist this twist and try to pull back.
221D—As he pulls back, reverse your action, step across and outside of leg. Continue your action.
221E—Using your hip as a lever continue the pull away from opponent. This action, though not shown here, twists opponent's arms, one over the other and breaks his grip. If opponent attempts to hang on to rifle, jerk it sharply; this will pull him to the deck over your hip.
221F—Having broken opponent's grasp, jam rifle butt back into his face, neck, chest or solar plexus, or step away and get rifle in ready position.

Rifle Twist (continued)

221G—Opponent has bayoneted rifle, in position to thrust or fire.
221H—Bend at hips and strike rifle at muzzle outward. From this position carry on with above series from picture 221B.

Snap Bayoneted Rifle to Deck

222A—Opponent has bayoneted rifle in position for a close thrust.
222B—Grasp rifle barrel from top at muzzle and force outward and downward. Twist body to side.
222C—Snap bayonet to the deck with both hands and kick opponent in the groin.

223A

223B

Outside Turn (Rifle High in Back)

223A—Opponent has rifle high in back, arms are raised.

223B—Step back and turn outside quickly, at this instant bring your arm up under barrel in bend of elbow.

223C—Grasp breech of rifle with near hand, maintaining lock on barrel with other hand and arm.

223D—Kick to opponent's groin, shins or knee caps. He'll let go of rifle!

223C

223D

Outside Turn—Hiplock

224A—Opponent has rifle in back, your arms are raised over bead.

224B—Turn outward and step back quickly, striking rifle with elbow.

224C—Pass striking arm under rifle barrel and lock in the bend of elbow. Other hand grasps rifle over breech.

224D—Secure rifle firmly and force downward. Opponent counters move.

224E—Snap rifle butt up into opponent's arm pit.

224F—Step through, keep rifle butt in opponent's arm pit and throw, using hip and leg as a lever. jam rifle butt to opponent's body or get in a ready position. Kick opponent's body if you wish.

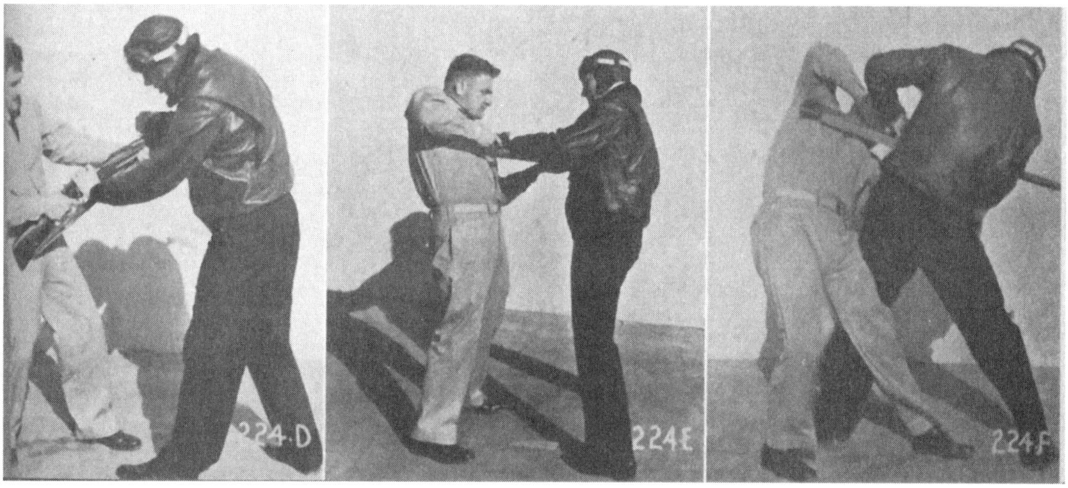

Inside Turn

225A—Opponent has rifle in back, your arms are raised over head.

225B—Step back and turn quickly inward. Note that this action gets you out of line of fire.

225C—Lock rifle barrel to side by grasping rifle breech and pinning barrel to side with elbow and forearm.

225D—Plant a kneelift in opponent's groin and attack face with free hand; gouge, or strike with the various hand blows listed in the fundamentals.

225E—Opponent has rifle with bayonet in back.

225F—Step back and twist inward as shown in illustration 225B. This gets you out of the way of any thrust made with the bayonet. Carry on from here as shown in illustrations 225C and 225D. The fact that the man with the bayoneted rifle must react to your move, makes it impossible for him to jab you with the bayonet.

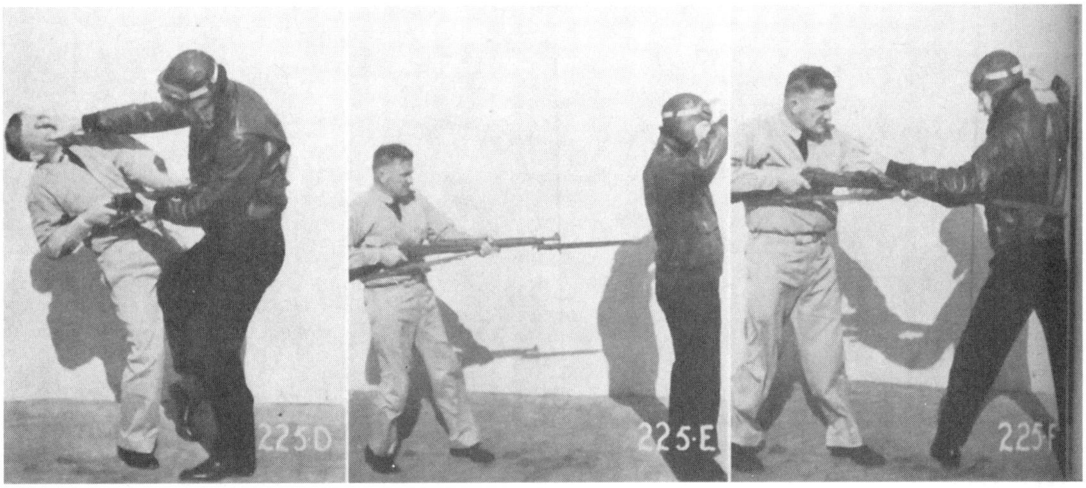

C. Club

(Or Any Weapon Used as a Club)

I. Unarmed

1. CROSS WRIST BLOCK—FLYING MARE (*See picture series 226*)

2. OUTWARD ARM PARRY (*See picture series 227*)

II. Armed With Club or Similar Weapon

1. CLUB PARRIES (*See picture series 228*)

In using a club as a defense weapon always keep hand end of club higher than the free end. This prevents assailant's blows from glancing down club to hand, striking knuckles or fingers.

2. CLUB STRIKES

These maneuvers will be discussed in Chapter XI, Part D.

Cross Wrist Block—Flying Mare

226A—Opponent attempts to strike at your head with a club or similar weapon. The natural inclination on your part would be to duck away or step back.

226B—Instead, you step in, crossing your arms above your head to ward off the blow. By stepping in, blow will pass harmlessly over your head and you are able to catch opponent's swinging arm with your upraised crossed arms.

226C—As opponent's arm contacts yours, turn quickly inward, grasping opponent's hand with one hand and securing his upper arm with the other. By catching opponent's hand with palm up you can lock his elbow. Snapping down sharply on shoulder will break elbow.

226D—Or drop to knee and pull opponent over head to the deck with a flying mare. The flying mare has been previously described. (See picture series 176)

Outward Arm Parry

227A—Opponent aims a blow at your head with a club.

227B—Step in, throw up arm and parry opponent's striking arm outward. Parry arm comes up inside striking arm.

227C—As blow intended for head passes harmlessly past shoulder, pass your parry arm over opponent's swinging arm.

227D—Lock arm with an elbowlock and attack groin with knee or deliver an edge of fist blow to jaw, or both. (Note elbowlock on far arm.) Other facial blows with the hands may be delivered from this position.

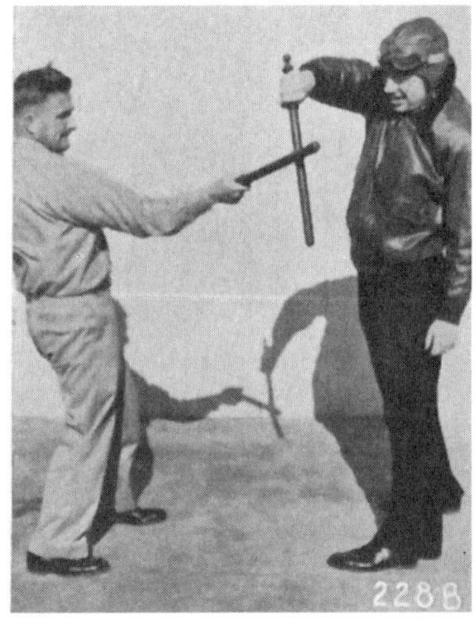

Club Parries

228A—To parry a downward blow, raise club overhead, hand end slightly higher than free end. Blow will slide harmlessly off free end.

228B—To parry a side blow, throw club arm across front of body, hand end high.

228C—To parry a low side blow, throw dub arm low across front of body, hand end high.

228D—To parry a low side back band, or a side back hand blow, throw club arm out, hand end of club held high.

D. Knives

(All Types)

I. Unarmed

1. DOWNWARD THRUST (*See picture series 229*)

2. SIDEWARD THRUST (*See picture series 230*)

3. UPWARD THRUST
    a. Outward Arm Twist, Hammerlock with Wrist Twist (*See picture series 231*)
    b. Inward Arm Twist (*See picture series 232*)
    c. Kick to Groin (*See picture series 233*)

II. Armed With Club or Similar Striking Weapon

1. PARRIES AND STRIKES (*See picture series 234*)

As in anything else we do, in order to become proficient, the above described maneuvers must be practiced repeatedly. It is not necessary to learn them all. You will discover that certain tactics will suit your needs better than others. Learn these to perfection, but nevertheless, have a working knowledge of the others. If slight variations are in order to suit your weight, size and height, include them as long as you do not destroy the effectiveness of the tactic.

Downward Thrust

229A—Opponent attempts a downward thrust with a knife or other sharp weapon. Although a downward thrust is generally reserved for the motion pictures, it would be well to know how to defend oneself against it.
229B—Reach up quickly blocking thrust with outstretched arm catching opponent's arm at the wrist. I weapon is short the forearm can be used to block thrust.
229C—This is a close-up of illustration 229B. Note that upraised thrusting arm is clasped at the wrist thumb downward. In other words, the entire hand catches the blow, not just the thumb as would be the case if the hand were held in a natural position.
229D—Retaining this grip on opponent's upraised arm, pass your other arm under upper part of his arm and over the forearm—
229E—Grasping your own wrist or opponent's wrist. Press away from your body locking opponent's elbow and shoulder.
229F—Close-up of 229E. Note position of your hands and how opponent's arm is securely locked. Pressure on arm will fracture elbow.

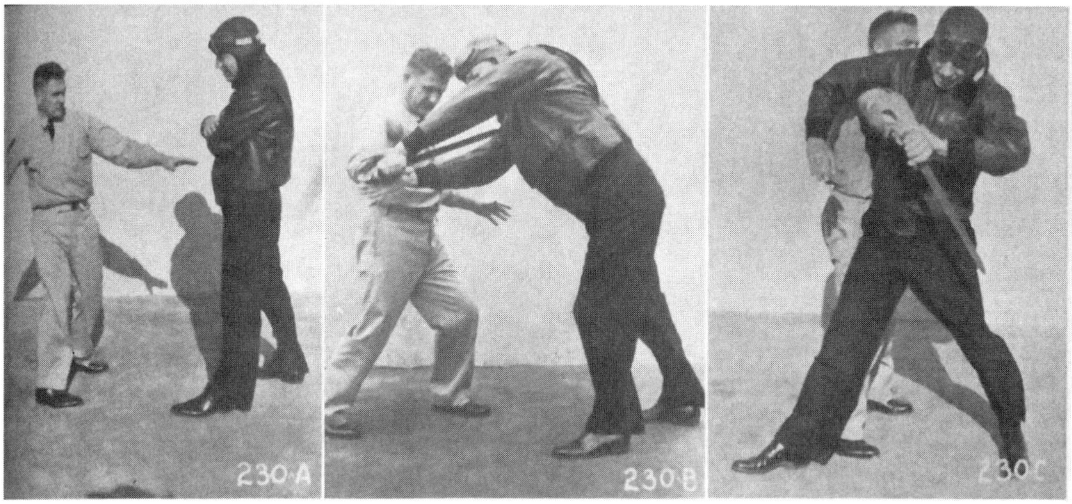

Sideward Thrust

230A—Opponent aims a side thrust with a knife at your body.

230B—Block thrust by forming a "V" with your hands (see illustration 231B). Keep arms straight.

230C—Clasp thrusting arm at wrist with one hand and pass other hand over upper arm above elbow, as you step into opponent.

230D—Apply a double wristlock, by grasping your own wrist. Note position of your hands.

230E—Step back and shove opponent's arm up his back into a twisting hammerlock. After securing opponent's wrist this maneuver is executed exactly the same as illustrated in picture series 143.

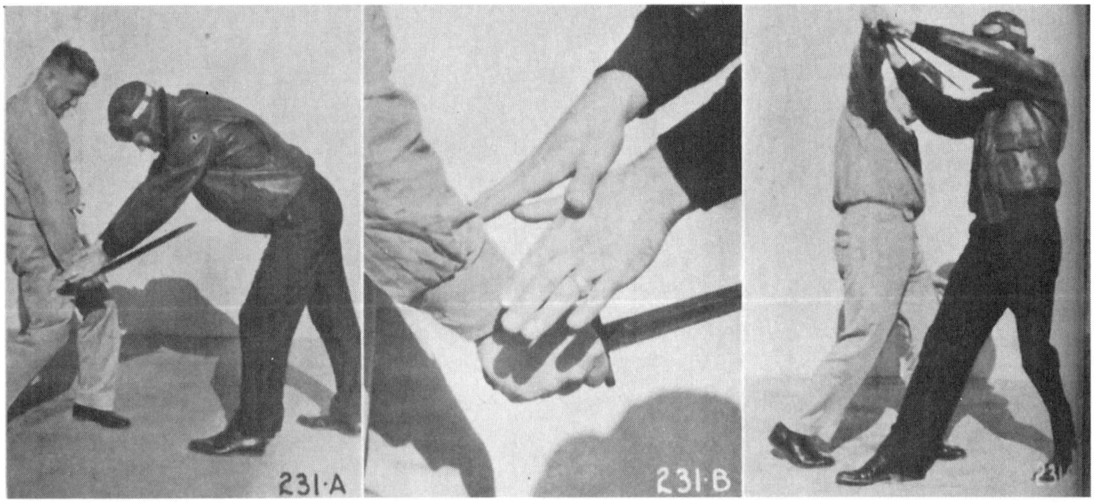

Upward Thrust—Outward Arm Twist, Hammerlock with Wrist Twist

231A—As opponent attempts an upward thrust, step in and block with a "V" formed with the hands Arms are outstretched and body is bent forward to keep as far from blade as possible.

231B—Close up of "V" formed with hands. This affords twice the catching area as does a single hand and protects the thumbs.

231C—Clasp opponent's wrist with both hands and jerk his arm upward sharply.

231D—Turn outward, pushing held arm over your head. Retain grasp.

231E—Now snap arm down, retain clasp on opponent's wrist with far hand, place near hand on elbow.

231F—Pull arm up opponent's back, press down on wrist and in on elbow. (This tactic, after wrist i secured, is executed in an identical manner as the maneuver illustrated in picture series 205.)

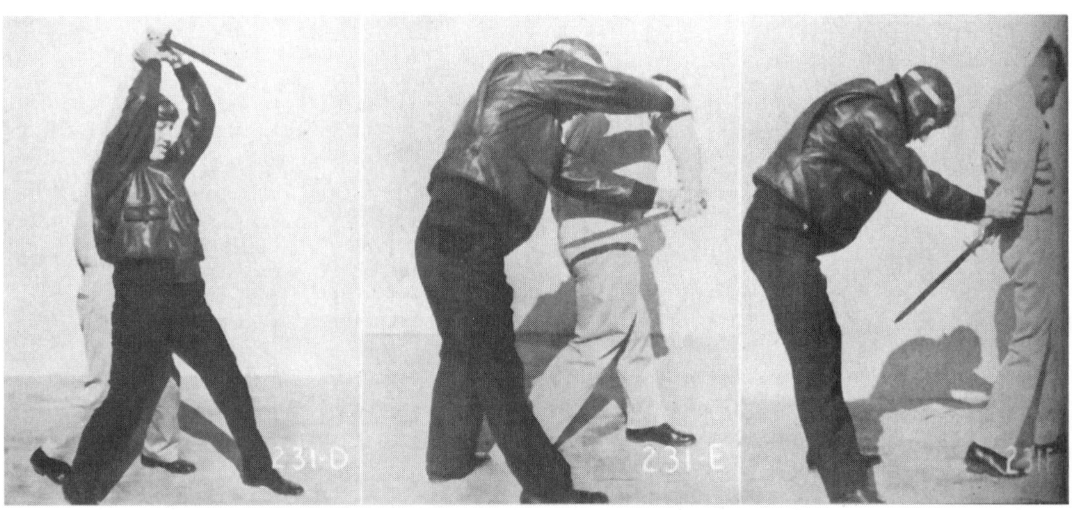

pward Thrust—Inward Arm Twist

32A—As opponent attempts an upward thrust, step in and secure striking wrist in a double hand "V."

32B—Snap opponent's arm up and begin an inward turn.

32C—Continue turn passing opponent's arm over head, retain grip on arm.

32D—As opponent's arm passes over head, his palm is up, thus the locked elbow may be snapped own over your shoulder breaking it. Also, opponent may be thrown over shoulder with a flying mare by nding at hips.

32E—However, if you wish to complete your turn, opponent's arm is locked with an outside double verse wristlock. Note how opponent's arm is bent over your inside arm; using it as a lever you can eak opponent's arm at the elbow or dislocate shoulder.

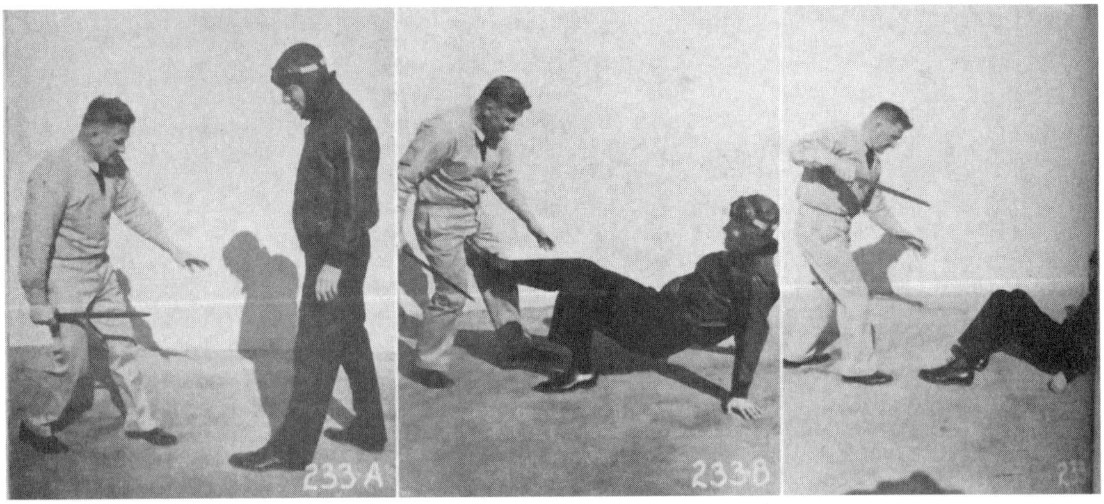

Upward Thrust-Kick to Groin

233A—Opponent aims an upward thrust of the knife at your stomach.
233B—Fall back out of knife thrust range and toe kick to groin. A straight thrust forward may be blocke
in the same manner.
233C—If you are on your back and an opponent attempts to attack with a knife thrust, kick to groin
use maneuver illustrated in picture series 163 (Toe Hook—Knee Kick).

Parries and Strikes

234A—As opponent delivers a downward thrust with a knife, step back and strike sharply with clu
blow to the inside of his forearm.
234B—As opponent delivers a side thrust, step back and strike sharply to the outside of his forearm.
234C—As opponent delivers an upward thrust, clasp both ends of club, bend body forward and parry b
striking downward against opponent's forearm. Other thrusts may be parried in a like manner.

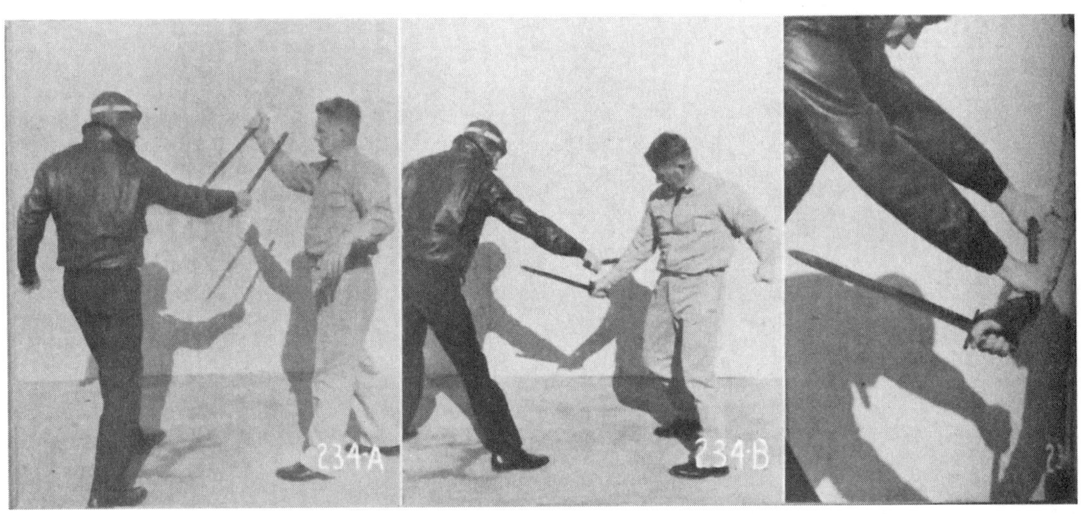

CHAPTER 11

Offensive Methods of Liquidating an Enemy

WHEN the critical moment arrives and it is either you or the other fellow in a life and death struggle, it is up to you to use the quickest and deadliest method of liquidating your enemy. Whether you be unarmed, armed with a club, knife, wire, gun, rock, bottle or other weapon, use your feet, hands or weapons to the greatest advantage.

This chapter illustrates those quick and easy methods not heretofore discussed, which for effectiveness and deadliness are without peer. In warfare, with all it grim possibilities, you must constantly remember, "He who fights fair is lost."

A. Attack from Rear

(Study maneuvers as listed referring to illustrations as needed.)

1. KICKS (*See illustrations 69, 70, 78, 79*)

The above mentioned kicks are extremely effective when attacking an opponent from the rear. A well-placed kick eliminates an enemy immediately, especially when speed is paramount.

2. HAND BLOWS (*See illustrations 16, 18, 19, 30, 31*)

Hand blows delivered to an opponent or enemy's back are very effective. Your victim does not know how you will attack or what part of his anatomy is vulnerable at the moment. He is unprepared in any case.

3. KNEELIFTS (*See illustration 55*)

Snap opponent's head back, arch his back and drive your knee up into his spine or kidneys.

4. EAR CLAP (*See picture series 235*)

5. LOCKED STRANGLES (*See illustrations 118 and 119*)

Follow instructions describing illustration 119 pertaining to forcing opponent to the deck. In this manner you will protect yourself against kicks and blows to the groin.

6. CLOTHING STRANGLES (*See picture series 236*)

7. FULL NELSON—FORWARD TRIP (*See picture series 237*)

8. LEG PICK—UP—DROP TO DECK
Approach opponent from rear, pick him up by legs at knees and shove forward to the deck. Keep your weight against back of legs for control.

Ear Clap
235A—Approach opponent from rear. Hands in position and cupped slightly.
235B—Bring hands together simultaneously over ears. This may break the ear drums, and will drop your opponent to the deck as a result of shock.

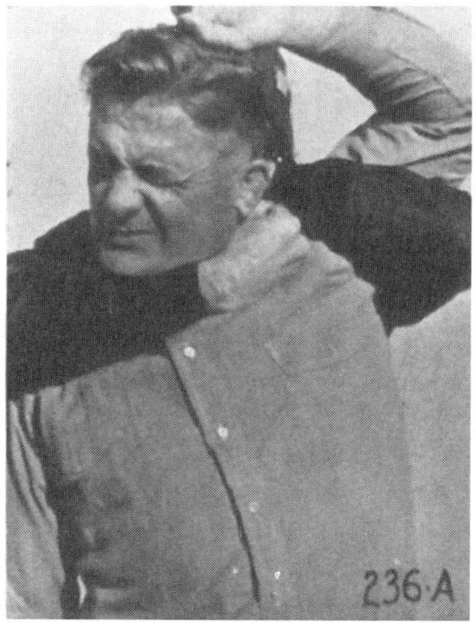

Clothing Strangles

236A—In this clothing strangle reach across front of opponent's throat and grasp collar, coat lapel, shirt or any loose clothing. Pull tight, pressing arm against throat while other arm passes under opponent's arm and behind neck in a half Nelson.

236B—Although this strangle is shown from the front, it can be applied equally as well from the rear. Just cross arms in front of opponent's throat, seizing clothing and press arms into the throat by pulling tight.

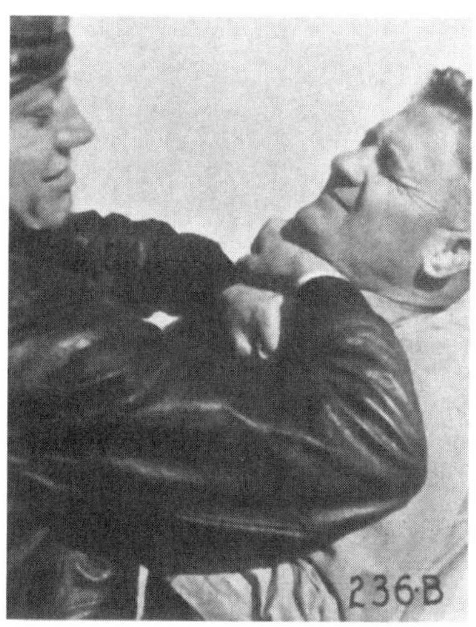

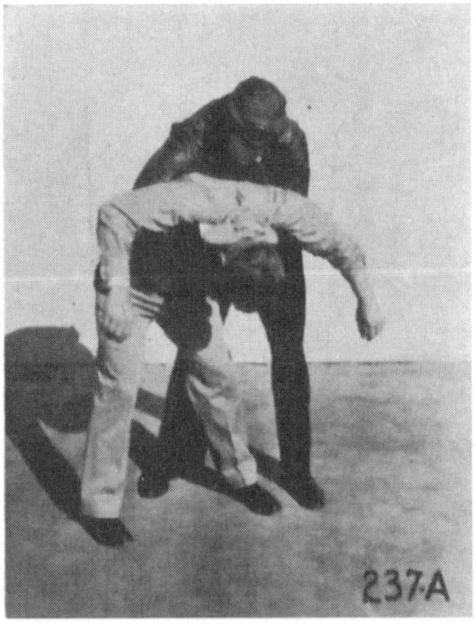

Full Nelson—Forward Trip

237A—Approach opponent from rear and bring your arms up under his arms and lock behind neck. This must be done quickly. Don't interlace fingers as shown in this illustration but lock them as shown in next illustration, 237B.

237B—Retaining the full Nelson, step quickly in front of opponent's legs and trip him on his head. He is unable to break his fall as his arms are locked. Using this trip with force will break his neck.

B. Wire or Cord Strangle

1. LOOP NECK—KNEE TO BACK (*See picture 238A*)

2. TWO HAND LOOP (*See picture series 239*)

The action shown in the three illustrations in series 239 takes only a fraction of a second to perform. The technique is well worth knowing. It is a silent form of liquidation from the rear. The pressure of the wire or strong cord on opponent's throat will prevent any outcry. A strand of piano wire or powerful fine garroting cord, approximately four feet in length is sufficient and can be easily stowed away in your clothing or concealed in the mouth. Each end of the wire or cord should be adjusted and protected with tape for your fingers. The sudden, severe jerk on the ends will cut deeply into an enemy's throat and end hostilities at once. This maneuver is very effective for ranger or commando work and when a sentry must be silenced or eliminated quickly.

3. BARBED WIRE ACROSS FACE

Approach enemy from rear and snap a strand of barbed wire across face (eyes, nose and mouth). Hold an end of wire in each hand and jerk back.

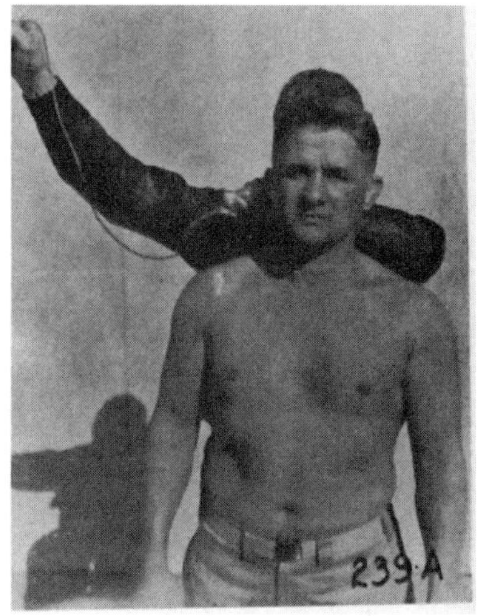

Loop Neck—Knee to Back

238A—Toss loop of wire or cord over neck of opponent from the rear, put knee in small of back and jerk with force. Continued pressure on wire or cord win strangle, the jerk may injure neck or spine. Quick application prevents opponent from making any outcry.

Two Hand Loop

239A—Note position of hands to apply loop to opponent's neck. One hand passes behind neck, or pushes wire or cord forward.

239B—Swing arm in front of throat looping wire.

239C—Complete loop and jerk arms sharply in opposite direction. Notice how wire or cord with little pressure digs into opponent's throat.

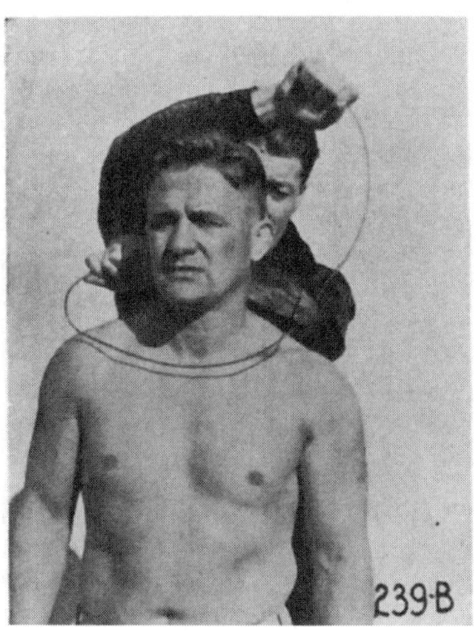

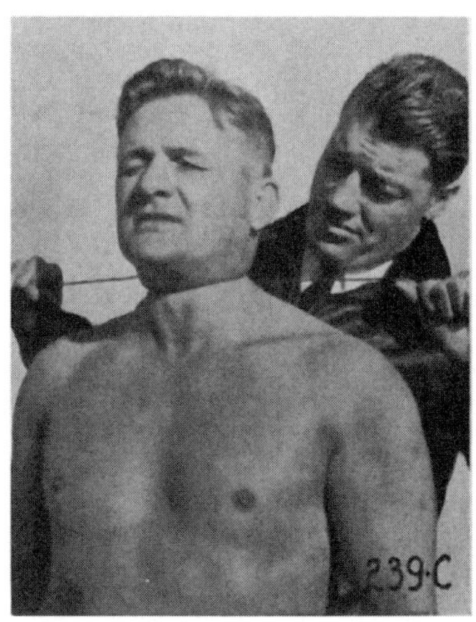

C. Knife

To kill an opponent easily with a knife, strike for any of the pressure points or vital areas vulnerable to attack at the moment. This may be a slice to an area where arteries are near the surface of the skin, a thrust or jab to deeper seated arteries and veins, a thrust or jab into the kidneys, stomach, eyes, ears, throat, etc., or a slice across muscles of the arms or legs to incapacitate the limb. Never try to jab at the heart, knife may become stuck between the ribs, strike rib, or glance from a rib.

An excellent means of using the knife offensively when attacking an enemy from the front is to strike out with a thrust directly from the hip. In other words, you do not use the down, side or up thrusts, but strike straight forward. This type of thrust is difficult to stop and offers you the greatest amount of protection.

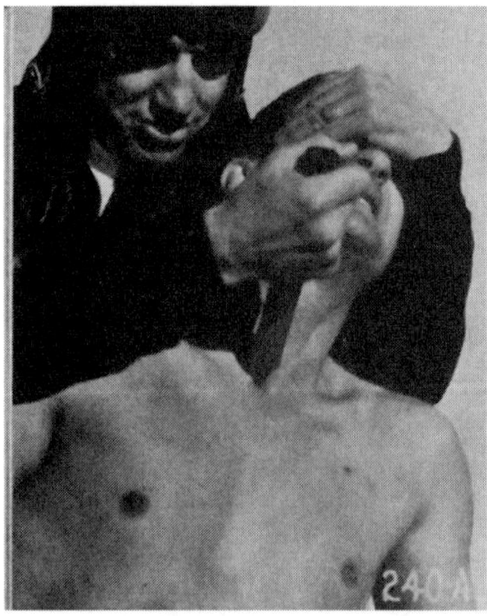

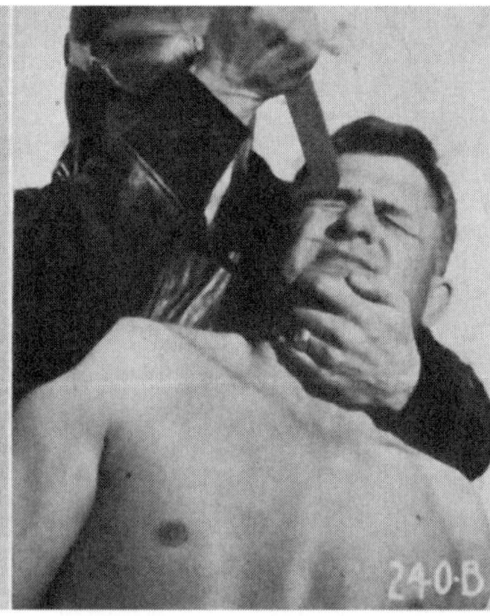

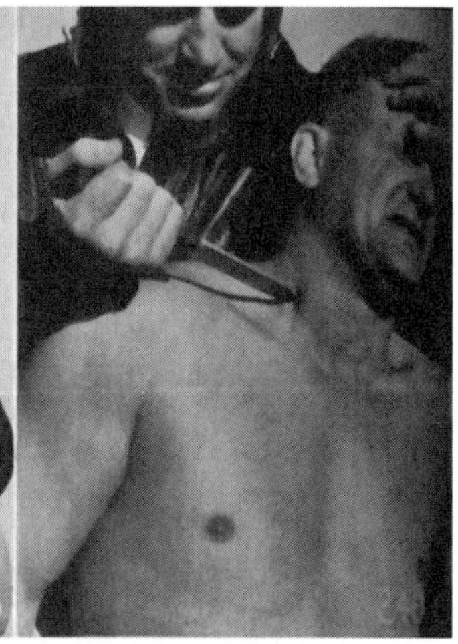

Jabs, Thrusts or Slices to Vulnerable or Vital Areas

240A—When behind, grasp opponent's head and thrust deeply into area at the base of the throat just above the juncture of the collar bones. Cuts windpipe and prevents outcry.

240B—In attacking from rear clasp chin or throat and thrust knife into eye. A sharp thrust will penetrate the brain providing knife has a narrow blade.

240C—Grasp opponent's head, twist to side and jab sharply into jugular vein and carotid artery.

240D—Or slice across side of neck cutting vein and artery.

240E—Attack from rear, throw arm about opponent's throat and thrust knife into kidneys. He will hemorrhage internally.

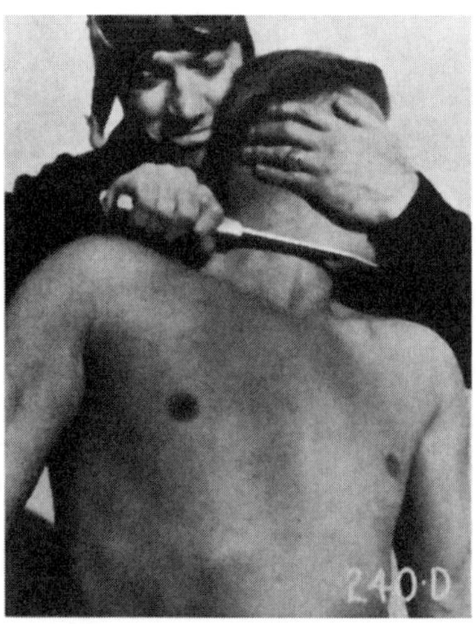

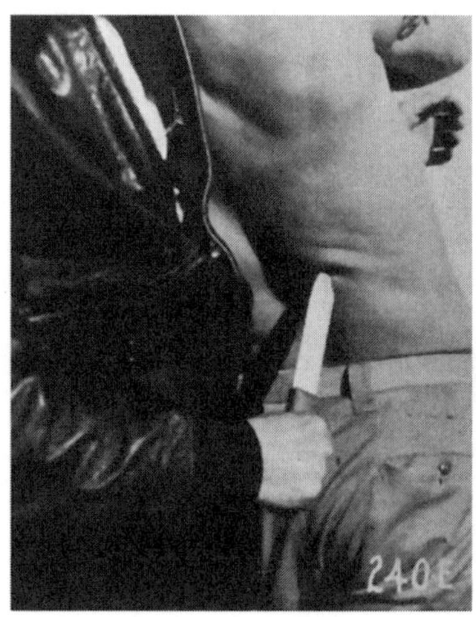

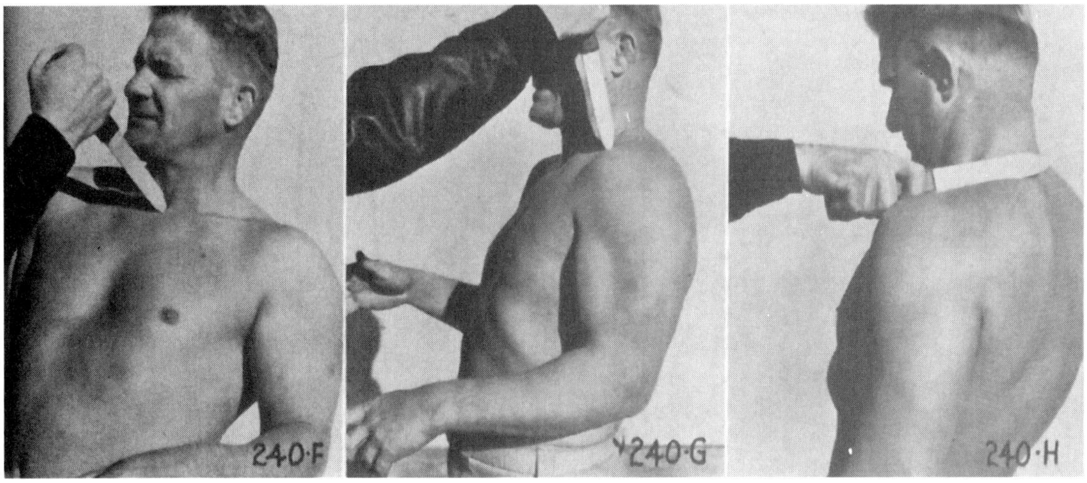

Jabs, Thrusts or Slices to Vulnerable or Vital Areas (continued)

240F—In attack from front jab knife deeply into area at base of throat just above juncture of collar bones. (See illustration 240A above.)

240G—Frontal attack, thrust down behind collar bone at side of neck, severing subclavical artery.

240H—Frontal attack, cut across muscles at side of neck, incapacitating arm.

240I—Frontal attack, jab or thrust upward under chin behind the point of jaw.

240J—Frontal attack, thrust deep into armpit. This will cut artery, veins and nerves and completely incapacitate arm.

240K—Frontal attack, thrust deep into stomach.

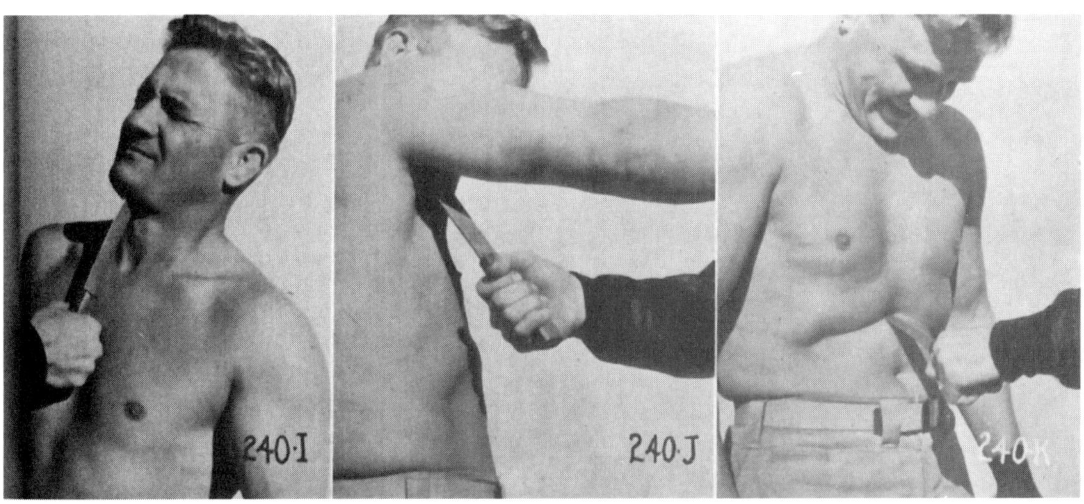

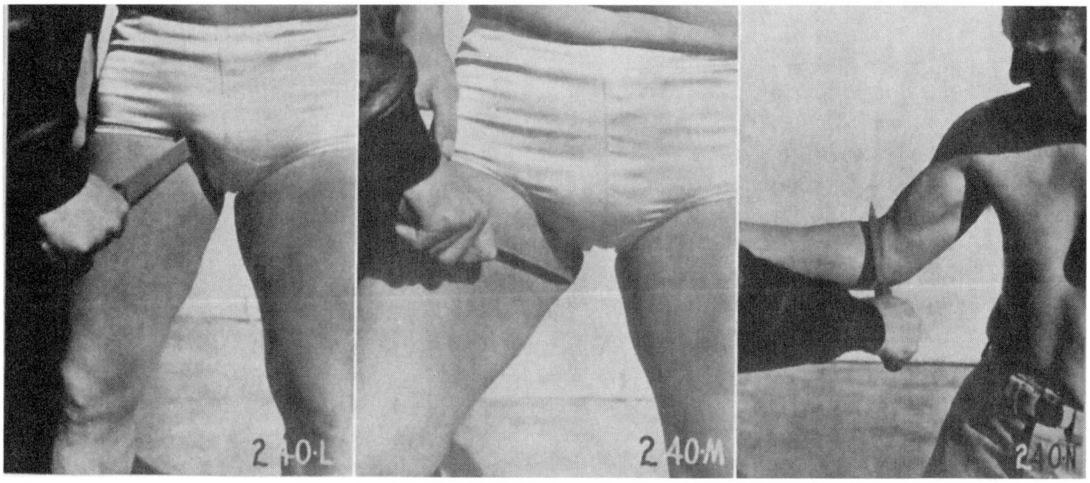

Jabs, Thrusts or Slices to Vulnerable or Vital Areas (continued)

240L—Frontal attack, jab knife into groin. This cuts arteries and veins exposed in this area.

240M—Frontal attack, slice to inside of upper leg at groin. Severs arteries and otherwise incapacitates limb.

240N—In a frontal attack, if opponent grasps you, slice across his arm at bend in the elbow. His arm will be useless for further action.

240O—Cut across inside of wrist if opponent clasps your arms or clothing.

240P—If in position, slice across cutting the Achilles' tendon. Foot will be useless thereafter.

240Q—It is possible to slice deeply into the back of the knee to stop resistance when you are behind an enemy. A jab into this area is also effective.

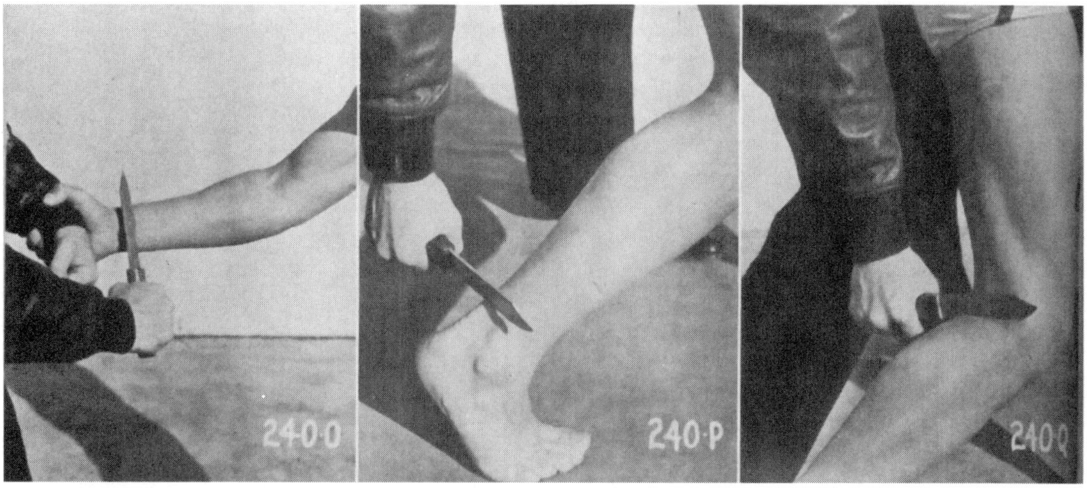

D. Club

1. MANEUVERS (*See picture series 241*)

A club or clublike weapon is undoubtedly an excellent defensive weapon and it can become a good offensive one as well. It can be used to thrust, strike or poke an opponent in vulnerable areas. Never make wild swings, be alert and deliberate. Not only can it be used offensively as illustrated, but it can be used to strike against the side of a knife, dagger, bayonet or sword, breaking and making them useless for further attack. When striking a sharp weapon don't strike the sharp edge, strike side of blade.

The club is a weapon which can be used in a great variety of ways. A man only needs to practice with it a short time before he realizes its possibilities. Remember that blows delivered by the ends of the club are better and more severe to your opponent than those from the side, primarily because of their greater penetrating effect.

The Japanese have used the billy club to subdue and kill unsuspecting sentries or other persons not wary enough to look behind them. Coming up behind their victim when they wished to do a silent job of eliminating, the club was thrown over the head and snapped sharply back against the larynx or Adam's apple with both hands, one at each end. This did not kill, but succeeded in paralyzing their victim's vocal cords so that he was unable to cry out in a warning or for aid. As the victim writhed in agony on the deck, it was a simple thing to drive a knife into his body.

A word of caution, however—don't become too complicated with club maneuvers. Simplicity of action in most cases is far more rapid than complex maneuvers.

A club when used as a bar (held by the hands at each end) becomes an excellent weapon to ward off any arm blows or kicks. (Note illustration 234C)

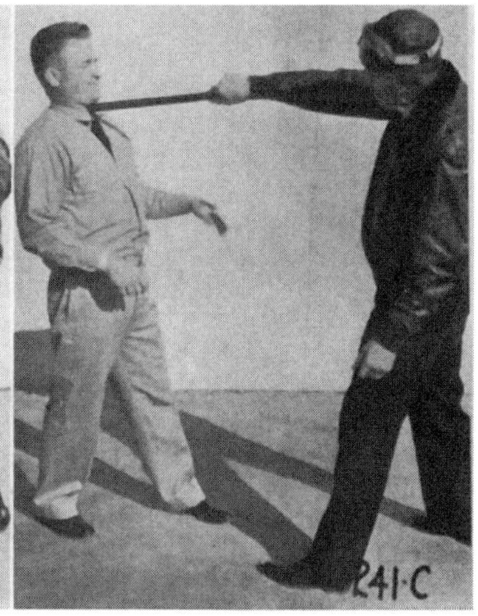

Club Maneuvers

241A—Using end of club, poke it into eyes of adversary, blinding him.

241B—In a like manner, poke end of club into opponent's mouth.

241C—A jab into an opponent's Adam's apple is very effective. The same type of jab to the solar plexus is also effective.

241D—Grasping club with both hands thrust it sharply into opponent's solar plexus

241E—Or directly up under the chin.

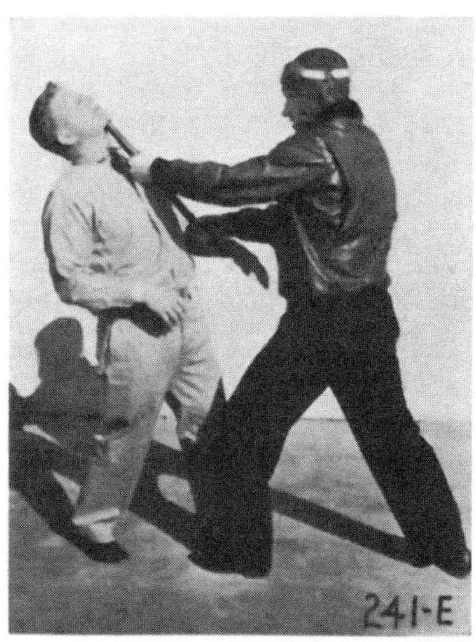

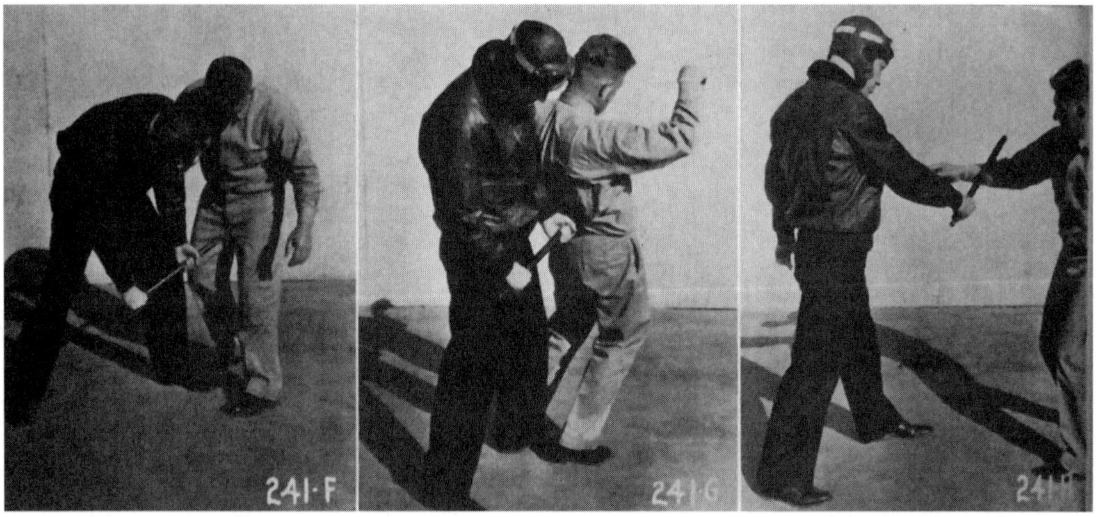

Club Maneuvers (continued)

241F—Thrust sharply into the groin using both hands for power.

241G—When in back of opponent, thrust sharply into his kidneys with the end of the club.

241H—Strike with club against forearm. (See picture series 234)

241I—A downward blow against side of opponent's jaw or across temple is an effective knockout blow.

241J—Striking down sharply to the collar bone will result in a fracture. A very good blow to deliver.

E. Opponent on Deck

When your opponent has been thrown or dropped to the deck, then is the time to use some quick subduing tactics. Use kicks, knee drops, blows to any vulnerable or vital parts of his body which are unprotected at the moment. Use all the locks and twists available for breaking bones, tearing muscles and joints loose from their sockets. Use your knife or club to the most effective points.

1. KICKS (*See illustrations 69, 71, 72, 80–82*)

Kick with force with the toe or heel to the head or neck, ribs, solar plexus. groin, spine or kidneys.

2. SHOULDER AND ARM LOCKS (*See illustrations 100, 101, 101B, 103, 107–113*)

It will be noted that several of the above illustrations show these arm and shoulder locks from a standing position. Nevertheless, all of these tactics may be applied either when opponent is on stomach or back. Note particularly illustration 113.

3. OTHER BONE—BREAKING AND JOINT FORCING MANEUVERS (*See illustrations 93–96, 114, 116, 126, 127, 127A*)

These additional maneuvers are effective under varying circumstances. Learn to use them where they will be most effective.

4. KNEE DROPS (*See illustrations 56–61*)

Drop on your opponent with the knee putting your weight behind it, to the neck, head, ribs, solar plexus, groin, spine and kidneys.

5. HAND BLOWS

All the various hand blows shown in the fundamentals can be used from various positions on the deck.

6. KNIFE (*See picture series 240*)

Use the thrusts, jabs and slices to the vital areas as illustrated.

F. Frontal Attack

It is never necessary to wait for your opponent to attack first. Carrying out any of the maneuvers listed in Chapter 6 can be done offensively. Go after your opponent! An offensive tactic not listed in the frontal attack chapter is the frontal leg pickup shown in picture series 242.

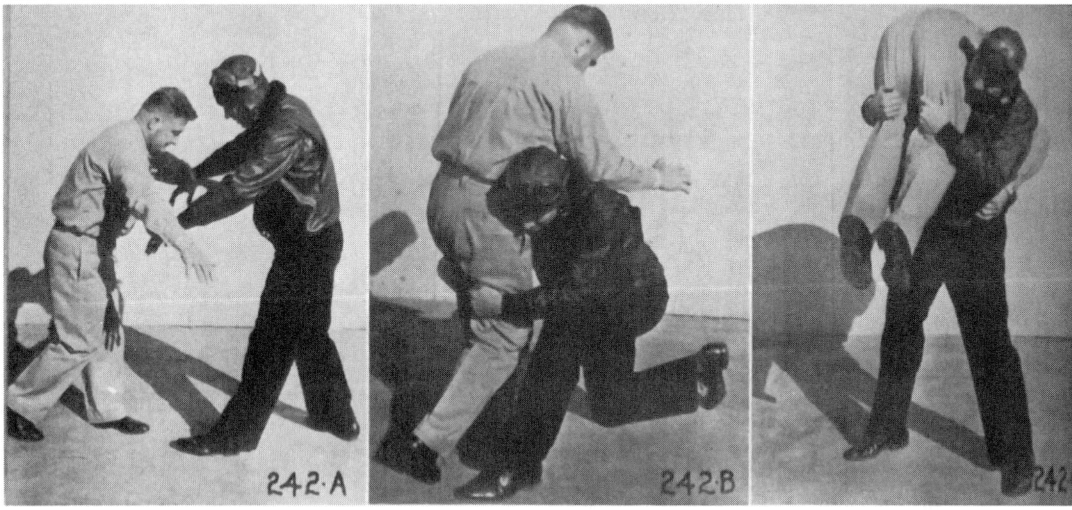

Frontal Leg Pickup

242A—Strike opponent's arms out to side. The strike is delivered inside of opponent's arms, outward.

242B—Drop to one knee as you step in and grasp opponent about the knees firmly to prevent him kneeing you.

242C—Raise opponent from the deck over shoulder.

242D—Shift under hand to opponent's back and other hand around legs.

242E—Keep your balance and drop opponent to head or back of neck on deck. You can easily injure your opponent with this maneuver.

G. Offensive Use of Firearms

In the maneuvers, tactics and techniques listed in this manual you have, very often, taken advantage of your opponent's mistakes. Don't let him, now, benefit by yours. This is especially true while handling firearms in emergency situations. Keep your pistol at your side, not out from body, and stay out of reach of enemy's arms or legs. Shoot to kill, if necessary, at the sight of any false moves. If armed with a rifle, hold it properly, but don't get within arm's reach of your opponent. Hold rifle down at hip, out of prisoner's reach. It can be fired effectively from this position. (Techniques for firing or otherwise handling firearms are covered in gunnery school; these are the finest methods available.)

243A—Proper technique for pistol. Keep it at side. Keep prisoner away and in a helpless position. Don't approach him until you have him at a distinct disadvantage.

243B—Proper technique for rifle. Rifle at hip, out of arm or leg reach of opponent. Don't approach until he is obviously at a disadvantage. (See searching techniques, picture series 190–194.)

Appendix

The Naval Aviation Hand-to-Hand Combat Program

The following organization plan is offered as an aid to the instructor of Hand-to-Hand Combat Activities.

I. Objectives
II. Facilities
III. Class Organization
IV. Procedure of Teaching
V. Lesson Plans
    A. The Pre-Flight School Program
    B. The Flight Training Program
    C. Advanced Lessons in Hand-to-Hand Combat

I. Objectives

A. To teach the future naval aviator all of the types and methods of dose-in or Hand-to-Hand Combat.
B. To acquaint him with the anatomical structural weaknesses and the most vital areas subject to attack.
C. To teach skills and fundamental techniques of Hand-to-Hand Combat, with particular emphasis on subduing or killing an opponent.
D. To teach methods of counterattack when an opponent is armed with a pistol, rifle, knife or club.
E. To develop poise and confidence in own ability to overcome an opponent in close-in fighting.

II. Facilities

A. Any grass plot or sandy beach area for outdoor instruction.
B. Mats or pads for covering deck when working aboard ship or indoors. (If mats or pads are not available, restrict throwing maneuvers.)
C. A minimum area of 8′ x 8′ should be allowed for each two men.
D. Raised platform 10′ x 10′ with matted surface and at least 30″ high for demonstrating maneuvers to the class.
E. Walls, radiators, pipes and all projecting objects should be padded or covered.

F. Equipment
  1. Rifle—Any regulation rifle will serve the purpose, although a wooden dummy rifle may be used for class work. (One rifle should be issued for each two men.)
  2. Knife—A hard rubber or leather knife is the finest substitute, although a wooden model will do. (One knife for each two men.)
  3. Club—Any type of club will serve the purpose. (One club for each man.)
  4. Wire—A heavy insulated electrical wire will serve the purpose for class instructional needs. Light cord may be used as a substitute. (One wire strand for each two men.)
  5. Pistol—Both the pistol and automatic types should be used to acquaint the men with peculiarities of each. (Old discarded weapons are preferred to wooden dummies.) (One pistol to each two men.)

G. Gear—Regular gym gear will serve the purpose, although old shirts, or blouses may be worn for clothing hold maneuvers. Regular heelless rubber soled shoes should be required for indoor drills.

III. Class Organization

A. The platoon or group will fall-in on the command. The tall men to the right, lining up according to height and weight. The odd numbers will team up with the even numbers to their right.

B. The group is now ready to take their preliminary warm-up exercises. These may be varied from day to day. Calisthenics may be used one day and combat types of contests may be used the next; as Hand Wrestling, Indian Wrestling, Leg Wrestling, pulling and pushing games. Five to ten minutes should be allowed for a thorough warm up, followed by a five-minute review of preceding days work in Hand-to-Hand Combat Activities.

C. The class is now ready for work on the "Whistle"! The class will come to "Attention"!

D. The head instructor will demonstrate the maneuver, with particular emphasis on variations and details. This will always be demonstrated from a raised platform which will enable the participants to observe the entire subjects, from head to foot.

E. The class will then be allowed to work the maneuver either by count, or informally. The contestants should be warned never to work under full resistance, the defensive man should work at about 50% of his ability (too many of the blows and maneuvers are death dealing tactics). The important objective in class drill is to teach the fundamentals and skills with enough time to master the timing and coordination required for effective use.

F. During the entire class period the instructor will be present, as will his assistants, each of whom will have charge of a small group. It will be the duty of each assistant to aid and coach his assigned group.

G. The class should be informal enough to allow each member to work the various maneuvers to suit his body characteristics. But, the class should never be allowed to become lackadaisical or out-of-control. The head instructor should keep the work progressive and interesting, by adding new skills and maneuvers as the given work is perfected.

H. The instructors and assistants should always have a complimentary word for a job well done.

IV. Procedure

A. Assembly

1. Fall-in!
2. Attention!
3. Normal interval—Dress right; Dress!
4. Ready—Front!
5. Count off by two's!
6. Odd numbers two steps forward—March!
7. Odd numbers—About face!

B. The class is now ready and in position to start work.

C. On the "Whistle"! all men will come to attention and advance to the instructor's platform where they will pay strict attention to a demonstration and explanation of a maneuver.

D. On command, "Carry on"! the groups will return to their allotted areas and work on the maneuver.

E. The assistant instructors will be assigned to a group of 16 to 24 men; they will aid and assist in coaching their respective groups. These assistants will grade the men in their detail each week, turning the grades in to the head coach.

F. Time of Period

The time allotment for each lesson is determined at each stage of training. As this time allotment may change with the expansion of the training program the officer in charge is instructed to follow the time allotment in force at that time. In the advance stages of flight training no set time allotment is made as the time available will vary with conditions.

V. Lesson Plans

A. The Pre-Flight School Program

Lesson #1. (see Introduction and Chapter 1)

1. Introduction of Hand-to-Hand Combat as an activity
    - a. Scope of program
    - b. Aims of program
    - c. Principals of Hand-to-Hand Tactics
2. Demonstration (Use a member of the class for a subject)
    - a. Areas most susceptible to attack (Chapter 4)
        - (1) Counter-joint movements or locking joints
        - (2) Vital areas
        - (3) Pressure points
        - (4) Nerve shocks
    - b. Fundamentals (Chapter 5)
        - (1) Stance
        - (2) Hand as a weapon
            - (a) Finger poke
            - (b) Finger jab
            - (c) Knuckle jab
            - (d) Fist
            - (e) Outer edge of fist
            - (f) Edge of hand
            - (g) Heel of hand
            - (h) Grasping hand
        - (3) Elbow as weapon
        - (4) Head as weapon
        - (5) Knee as weapon
        - (6) Foot as weapon
        - (7) Heel as weapon
        - (8) Fingerlock
        - (9) Handlock
        - (10) Wristlocks
        - (11) Elbowlocks
        - (12) Hammerlock
        - (13) Shoulderlock
        - (14) Necklock
        - (15) Hiplocks

(16) Leglocks
(17) Flying mare
(18) Nerve shocks as weapons

Lesson #2. Demonstration and Practice

1. Frontal Attack (Chapter 6)
    a. Rushing
        (1) Kicks
        (2) Hand blows
        (3) Head twist chancery
        (4) Chancery strangle
        (5) Arm drag
        (6) Switch
        (7) Double wristlock
    b. Pushing or striking
        (1) Kicks
        (2) Hand blows
        (3) Elbow blows
        (4) Kneelift
        (5) Handlocks
        (6) Fingerlocks
        (7) Regular wristlock
        (8) Reverse wristlock
        (9) Elbowlocks
            (a) Single elbowlock and leg trip
            (b) Double elbowlock and kicks
            (c) Twisting elbowlock
    c. Grasping
        (1) Kicks
        (2) Hand blows
        (3) Elbow blows
        (4) Kneelift
        (5) Release
        (6) Regular wristlock
        (7) Reverse wristlock
        (8) Roll against elbow (outboard)
        (9) Double wristlock
        (10) Fingerlocks
        (11) Regular hiplock
        (12) Reverse hiplock
    d. Strangling (Arms-length)
        (1) Kicks
        (2) Hand blows

(3) Elbow blows
(4) Grasping hand
(5) Nerve shocks
(6) Fingerlocks
(7) Regular wristlock
(8) Reverse wristlock
(9) Elbowlocks
(10) Outside blow to locked elbow
(11) Regular hiplock
(12) Reverse hiplock
(13) Strangle hiplock (arm wedge)
(14) Arm wedge
(a) Hand blows
(b) Elbow blows
(c) Kicks
(15) Finger pokes

Lesson #3. Demonstration and Practice
1. Frontal Attack (continued) Chapter 6
a. Strangle (closed)
(1) Kicks
(2) Hand blows
(3) Grasping hand
(4) Head snap backwards
(5) Nerve shocks—vital areas
(6) Head butts
(7) Bite jugular, neck or ear
b. Body locking—(Bearhug) Arms included
(1) Kicks
(2) Hand blows
(3) Groin grasp
(4) Head butts
(5) Kneelifts
(6) Bites
(7) Trips
(8) Hiplocks
(9) Head snap backwards

c. Body Locking—Arms free
  (1) Kicks
  (2) Hand blows
  (3) Nerve shocks
  (4) Elbow blows
  (5) Kneelifts
  (6) Neck snap backwards
  (7) Leg trip
  (8) Clap over ears
  (9) Regular hiplock
  (10) Reverse hiplock
  (11) Double wristlock
  (12) Elbowlocks

Lesson #4. Demonstration and Practice
1. Rear Attack (Chapter 8)
  a. Striking or pushing
    (1) Edge of band or fist blows
    (2) Heel kicks
    (3) Single or double mule kick
    (4) Turn and hiplock
  b. Grasping
    (1) Kicks
    (2) Hand blows
    (3) Elbow blows
    (4) Twisting elbowlock
    (5) Double elbowlock
    (6) Regular wristlock } (from hair or neck grasp)
    (7) Reverse wristlock } (from hair or neck grasp)
    (8) Fingerlocks } (from hair or neck grasp)
  c. Strangling (arms-length)
    (1) Kicks
    (2) Hand blows
    (3) Elbow blows
    (4) Duck and turn
    (5) Grasp both thumbs, duck and turn, force elbows to lock
    (6) Grasp thumb, turn palm up, drive shoulder into elbow
    (7) Turn outward against elbow

Lesson #5. Demonstration and Practice

1. Rear Attack continued (Chapter 8)
    a. Strangling (closed)
        (1) Kicks
        (2) Hand blows
        (3) Grasp to groin
        (4) Elbow blows
        (5) Head butts
        (6) Flying mare
            (a) Drop to knee
            (b) Throw from hips
            (c) Hair grasp
            (d) Locked elbow or elbows
        (7) Grasp elbow, push up to hammerlock (swinging release
        (8) Twist to side, step to rear, trip backward
        (9) Reverse wristlock
        (10) Reverse double wristlock

    b. Bodylocking (arms included)
        (1) Kicks
        (2) Head butts
        (3) Elbow blows
        (4) Groin grasp or strike
        (5) Fingerlocks
        (6) Squat, raise arms out, turning, go behind, grapevine arm
    c. Bodylocking (arms free)
        (1) Kicks
        (2) Head butts
        (3) Knuckle strike
        (4) Elbow blows
        (5) Groin grasp or strike
        (6) Fingerlocks
        (7) Double wristlock
        (8) Rear hiplock
        (9) Double wristlock, and hiplock
        (10) Leg pick up—between legs
        (11) Leg pick up—outside legs, (single)
        (12) Leg pick up—outside legs, (double)
        (13) Switch

Lesson #6. Demonstration and Practice

1. Kicking (Chapter 7)
    a. Upright
        (1) Avoiding kicks
        (2) Counter-kick or trip and blows
        (3) Double cross-arm catch
        (4) Step inside, kneelift, heel of hand to jaw, etc.
        (5) Grasp leg—toe hold inward, bar toe hold—attack from rear
    b. From the deck
        (1) Toe hook on heel, kick to knee
        (2) Scissor kicks to bar toe holds
        (3) Catch kicking leg, groin kick
        (4) Low body block to twisting kneelock
        (5) Mule kick; single or double kick with free leg if one is caught
        (6) Roll out of range, counter kick

Lesson # 7. Demonstration and Practice

1. Club and Knife Maneuvers (Chapter 10)
    a. Club
        (1) Unarmed
            (a) Cross-wrist block, flying mare
            (b) Outward arm party-attack body or elbow block
        (2) Armed with club or similar weapon
            (a) Club parries
            (b) Outward arm parry-attack body or elbow block
    b. Knife
        (1) Unarmed
            (a) Upward thrust
            (b) Sideward thrust
            (c) Downward thrust
        (2) Armed with club of similar weapon
            (a) Parries
            (b) Strikes

Lesson #8. Demonstration and Practice

1. Pistol and Rifle Maneuvers (Chapter 10)
    a. Pistol (Lecture on peculiarities of)
        (1) Shoulder or cross draw holster

(a) Reverse wristlock
(2) Side holster
(a) Twisting hammerlock
(3) Gun drawn (hands raised)
(a) Scissor swing
(4) Gun drawn (hands down)
(a) Up and down scissors
(5) Gun in back
(a) Outside turn
(b) Inside turn
b. Rifle
(1) Facing opponent
(a) Rifle twist
(b) Rifle snap
(2) Rifle in back
(a) Inside turn

Lesson #9. Demonstration and Practice
1. Offensive methods of liquidating an enemy (Chapter 11)
a. Attacks from rear
(1) Kicks
(2) Kneelifts
(3) Ear clap
(4) Locked strangles
(5) Clothing strangles
(6) Full Nelson
b. Wire or cord strangles
(1) Loop to neck, knee to back jerk
(2) Two hand loop, jerk
(3) Barbed-wire across face
c. Knife
(1) Jabs, thrusts, and slices to vulnerable areas, pressure points
d. Club
(1) Club maneuvers
e. Opponent on deck
(1) Kicks
(2) Shoulder and armlocks
(3) Knee drops
(4) Hand blows

(5) Knife
f. Proper use of firearms
(1) Pistol
(2) Rifle

Lesson # 10.

1. Show movie of Hand-to-Hand Combat
2. If movie is not available, use this period for general review of highlights of first nine drills.

B. Flight Training Program

Introduction: The Hand-to-Hand Combat program now being given at the pre-flight schools is a new activity for the vast majority of the cadets. True, some cadets may have had some lessons or practice in jui jitsu, judo, wrestling, boxing or fencing which are allied activities of Hand-to-Hand Combat. But on the whole the maneuvers of Hand-to-Hand will leave the cadet wanting for more lessons in the fundamentals.

The present set-up of lessons in the pre-flight schools allows for 10 periods. This certainly is not time enough to allow for perfection of all maneuvers listed for the pre-flight schools. With this thought in mind additional drills for all stages of flight training have been set up, allowing time for review of lessons given at the preflight school. The main objective should be to perfect timing of all maneuvers with emphasis on variations. The lessons following are more detailed than the ones used at the preflight school.

The instructor should run through the preliminary lessons permitting most of the class time for detailed study and practice of the advanced combinations.

The class organization must cover all the maneuvers listed. Particular emphasis should be given to the various situations that arise in actual warfare. For example: in the pre-flight school men were paired off according to weight, and height. Now, the subjects may be mixed up. A tall man working with a short man which will change the situation regarding some maneuvers, then the light weight should work with a heavier boy to learn bow to offset weight with speed and variations of maneuvers to get his opponent out of range of effectiveness. These problems come up in warfare. Now is the time to be introduced to them. Put emphasis on the ingenuity of each boy thinking and acting for himself.

Particular emphasis should be put on new disarming maneuvers which have been added. Since the basic combat maneuvers predominated in the pre-flight school lessons, only one maneuver was given for each position or situation. The disarming lessons are reviewed with several variations added for each position.

Lesson #11. Demonstration and Practice

1. Review of Pre-Flight School lessons
    - a. Fundamentals
    - b. Frontal and Rear attacks
    - c. Kicking maneuvers
    - d. Disarming maneuvers against club, knife, pistol and rifle
    - e. Offensive methods of liquidating an opponent

Lesson #12. Demonstration and Practice

1. Frontal attack (Chapter 6)
    - a. The following situations should be reviewed with emphasis on the most effective counterattack for each: Rushing, pushing, striking, grasping, strangle-arms length, bodylock—arms included, and bodylock—arms free.
    - b. Counterattacks used in pre-flight schools
        - (1) Kicks
        - (2) Head chancery, strangle or twisting
        - (3) Double wristlock
        - (4) Hand blows or grasping
        - (5) Regular wristlocks
        - (6) Reverse wristlocks
        - (7) Handlocks;
        - (8) Regular hiplocks
        - (9) Reverse hiplocks
        - (10) Nerve shocks
        - (11) Releases
        - (12) Back flip and kick
        - (13) Elbow blows
        - (14) Switch
        - (15) Arm drag

(16) Kneelifts
(17) Head butts
(18) Neck snap
(19) Clap over ears
(20) Arm wedge blows

Lesson #13. Demonstration and Practice

1. Rear Attack (Chapter 8)
    a. The following situations should be reviewed with emphasis on the most effective counter-attack for each: Pushing, striking, grasping, strangle—arms length, strangle—closed, bodylocking—arms included, bodylocking—arms free.
    b. Counterattacks used in pre-flight schools
        (1) Kicks
        (2) Hand and elbow blows
        (3) Fingerlocks
        (4) Regular wristlocks;
        (5) Reverse wristlocks
        (6) Double wristlocks
        (7) Flying mare
        (8) Groin grasp or strike
        (9) Elbow locks
        (10) Head butts
        (11) Regular hiplock
        (12) Reverse hiplock
        (13) Nerve shocks
        (14) Leg pick up and trips (single and double)
        (15) Switch
        (16) Twist to side, step to rear, trip backwards
        (17) Single and double, mule kicks
        (18) Leg trip

Lesson #14. Demonstration and Practice

1. Kicking (Chapter 7)
    a. Upright
        (1) Avoid kicks
        (2) Counter-kicks or trip, close range and use blows
        (3) Cross-arm catch

(4) Step inside, kneelift, heel of hand to jaw, etc.
(5) Grasp leg, toe turn inward, bar toe hold attack from rear

b. From the deck
(1) Toe hook on heel, knee kick
(2) Scissor kicks to bar toe hold
(3) Grasping kicking leg, groin kicks
(4) Low body block to twisting kneelock
(5) Mule kick; kick with free foot if one is caught
(6) Roll from range, counter kick

Lesson #15. Demonstration and Practice

1. Offensive attacks to subdue or kill an enemy (Chapter 11)

a. Attacking from rear
(1) Kicks
(2) Hand blows
(3) Kneelifts,
(4) Ear clap
(5) Locked strangles
(6) Clothing strangles
(7) Full Nelson
(8) Leg pick up—trip to deck

b. Wire or cord strangle
(1) Loop to neck, jerk, kneelift to back
(2) Two handed loop, jerk
(3) Barbed wire across face, jerk

c. Knife
(1) jabs, thrusts, and slices to vulnerable areas, and pressure points

d. Club
(1) Offensive club maneuver

e. Opponent on Deck
(1) Kicks
(2) Knee drops
(3) Shoulder and armlocks
(4) Hand blows
(5) Knife

f. Proper use of firearms
(1) Pistol
(2) Rifle

Lesson #16. Demonstration and Practice
1. Disarming maneuvers against club, knife (Chapter 10)
a. Club
(1) Unarmed
(a) Cross wrist block, flying mare
(b) Outward arm parry
(2) Armed with club or similar weapon
(a) Club parries
(3) Offensive club maneuvers
b. Knife
(1) Unarmed
(a) Upward thrust
(b) Sideward thrust
(c) Downward thrust
(2) Armed with club or similar weapon
(a) Parries

Lesson #17. Demonstration and Practice
1. Disarming maneuvers against a rifle (rifle bayoneted or unbayoneted) (Chapter 10)
a. Facing opponent
(1) Rifle twist
(2) Rifle strike and grasp, frontal attack
(3) Rifle strike, snap bayoneted rifle into ground
(4) Rifle snap
b. Rifle in back
(1) Outside turn, lock rifle and attack
(2) Inside turn, lock rifle and attack
(3) Rifle twist from elbow strike
(4) Outside turn, hiplock throw
c. Work on variations of above attacks as: rifle pointed high into face, rifle pointed into chest, rifle pointed into stomach, or same maneuvers from the rear.

Lesson #18. Demonstration and Practice
1. Disarming maneuvers against pistol (Lecture 5 to 10 min.

on peculiarities of small firearms; pistol and automatic.) (Chapter 10)

a. Pistol
- (1) Shoulder or cross draw holster
  - (a) Reverse wristlock
  - (b) Face strike with pistol
- (2) Side holster
  - (a) Twisting hammerlock
  - (b) Double wristlock
- (3) Gun drawn (hands raised)
  - (a) Inside turn
  - (b) Outside turn
  - (c) Scissor swing (gun in face)
  - (d) Scissor swing (gun in stomach)
  - (e) Emergency—gun at side, shoulder grasp, reverse wristlock
- (4) Gun drawn (hands down)
  - (a) Up and down scissors
  - (b) Inside turn
  - (c) Outside turn
  - (d) Arm strike (gun at temple)
- (5) Gun in back
  - (a) Outside turn
  - (b) Inside turn

Lesson #19. Demonstration and Practice

1. Searching and techniques for control and leading of prisoners (Chapter 9)

a. Methods of search
- (1) On stomach, arms over head
- (2) Kneeling, arms behind back or neck
- (3) Wall method
- (4) Upright, arms high over head
- (5) Prisoners feigning death
  - (a) Knee in neck or back
  - (b) Bar toe hold
  - (c) Double bar toe hold

b. Controlling and leading
- (1) Cross wrist drag

(2) Cross arm drag
(3) Hock of knee snap
(4) Handshake to cross chest or neck lead
(5) Fingerlocks
(6) Fingerlocks with elbowlock or tuck
(7) Wristlock with elbow tuck
(8) Wristlock with mouth lead
(9) Elbowlock and twist wrist to opposite arm
(10) Outward arm twist, hammerlock, wrist twist
(11) Cuff grasp, to hammerlock

Lesson #20. Demonstration and Practice

1. This entire period should be devoted to a review of the highlights of each situation as studied in the preceding drills. Particular emphasis should be placed on the disarming lessons since several maneuvers had been added to these series.

C. Advanced Lessons in Hand-to-Hand Combat

The following lessons are set up to give the personnel a physical work-out while reviewing the various Hand-to-Hand Combat maneuvers. Emphasis now will be placed on using a particular hold from all the various positions and situations with the effect, timing and execution being the main objectives. The class may practice these maneuvers and holds at about 50% resistance. The Instructor may introduce the "Contest Type" of drill in which both men will attempt to secure an advantageous position, hold or maneuver (similar to "controlled boxing drills). A running score for the number of successful attempts will stimulate interest and a competitive spirit. During these drills both men must stay within range of each other, one being on the offensive and the other on the defensive. All maneuvers should be given a time limit of a minute or so. The timing element can be used for the entire group of participants, starting action on a signal and stopping on a signal or bell.

Note: Again attention is called to the fact that many of the maneuvers cannot be executed under actual combat conditions, because of their damaging results. It is therefore the instructor's duty to keep the group under control. Although, in some instances, as when working the Club Maneuvers the contestants may use football helmets and shoulder pads to guard against damaging blows which would permit the men to work under actual combat conditions.

Lesson #21.

The first lesson should include the showing of the U. S. Navy Training Film "Hand-to-Hand Combat" this will act as a refresher and prepare the men to work each maneuver through as a unit, even though they may use several holds or maneuvers to achieve the proper result.

If the film is not available, then the instructor takes up the time by lecturing and demonstrating the various positions and situations of Hand-to-Hand Combat with particular emphasis on the two or three "best" holds, blows, kicks, or throws for each position.

Lesson #22. Review use of various weapons (Chapter 5)

1. Hand blows
    a. Edge of hand or edge of fist
        (1) Across bridge of nose
        (2) Across upper lip
        (3) Temple
        (4) Side of jaw
        (5) Side of neck
        (6) Collar bone
        (7) Front of neck
        (8) Solar plexus
        (9) Groin
        (10) Base of skull
    b. Fist
        (1) Solar plexus
        (2) Bridge of nose
        (3) Kidney punch
        (4) Groin
        (5) Base of skull
    c. Fingers
        (1) Eye gouge
        (2) Throat thrust or jab
        (3) Solar plexus

(4) Groin
(5) Mouth rip
(6) Nostril rip

d. Heel of hand
(1) To jaw
(2) To nose

e. Knuckle jab
(1) Nose
(2) Eyes
(3) Upper lip
(4) Adam's apple
(5) Solar plexus
(6) Back of closed fist (metacarpus)

f. Grasping hand
(1) All nerve shocks—especially the vagus; nerve in neck
(2) Jugular vein and carotid artery
(3) Throat
(4) Lips
(5) Ears

2. Elbow as weapon

a. Frontal attack
(1) Up to chin
(2) Down to chin
(3) Front lateral to face or chin
(4) Back lateral to face or chin
(5) Down to collar bone

b. Rear attack
(1) Snap to face or chin
(2) Snap to solar plexus or ribs
(3) Snap to groin

3. Knee as weapon

a. Kneelift to or knee drop to:
(1) Face
(2) Neck
(3) Chest
(4) Stomach
(5) Groin
(6) Back (spine or kidneys)

4. Foot as weapon
    a. Toe kicks
        (1) Shin
        (2) Kneecap
        (3) Groin
        (4) Stomach
        (5) Chest
        (6) Neck and head
        (7) Hock of knee
        (8) Coccyx (tail bone)
        (9) Spine (especially around head and neck
    b. Heel kicks
        (1) Stomp to arch of foot
        (2) Shin
        (3) Kneecap
        (4) Groin
        (5) Ribs—stomach
        (6) Face
        (7) Neck and head
        (8) Small of Back

Lesson #23. Methods of locking the various joints of the body (Chapter 5)

1. Fingerlock (or fingers)
2. Wristlock
    a. Regular
    b. Reverse
    c. Double wristlock
3. Elbowlock
    a. Single armlock
    b. Double armlock
    c. Twisting lock
4. Shoulderlock
5. Neck snap or necklock
    a. Neck snap backward
    b. Full Nelson lock
    c. Chancery stranglelock, front and rear
6. Leg lock (Twists)
    a. Inward turn

b. Outward turn

7. Kneelock

a. Force of kick to knee

b. Bar toe hold

c. Double bar toe hold

Lesson #24. Searching and techniques for controlling and leading (Chapter 9)

1. Methods of search

a. Kneeling, arms behind back

b. On stomach, arms over head

c. Upright, arms high

d. Wall method

e. Prisoner on deck, may be feinting death

(1) Knee in neck or back

(2) Bar toe hold

(3) Double bar toe hold

f. Safety measures in taking prisoners off

(1) Arms well extended over head or clasped behind neck

(2) Cut belt or suspenders, drop pants to half mast

(3) Jerk coat collar down to elbows of unarmed prisoners

(4) Gun always close at side

(5) If two men are searching, keep out of each other's line of fire

(6) Turning corner (one or more prisoners) step well out to side to hold clear view on all men, never let one out of sight

2. Controlling and leading

a. Cross wrist drag

b. Cross arm drag

c. Hock of knee snap

d. Handshake, to cross chest or neck lead

e. Fingerlocks, with arm or elbow tucks

f. Wristlock with elbow tucked

g. Reverse wristlock and belt grasp

h. Wristlock and mouth lead

i. Elbowlock and wrist twist to opposite arm

j. Outward arm twist, hammerlock, wrist
k. Twist
l. Cuff grasp to hammerlock

Lesson #25. Throwing maneuvers
1. From frontal attack (Chapter 6)
   a. Regular hiplock
   b. Reverse hiplock
   c. Strangle hiplock
   d. Single elbowlock and leg trip
   e. Double wristlock and hiplock
   f. Back flip
2. From rear attack (Chapter 8)
   a. Hair grasp, flying mare
   b. Head grasp, flying mare
   c. Elbow grasp, flying mare
   d. Palm uppermost, elbowlock, flying mare
   e. Single leg pick up, between own legs, sit back against knee
   f. Single leg pick up, step to side and trip free leg, drop opponent to deck
   g. Step to side, double pick up, drop opponent on head
   h. Leg trip
   i. Reverse hiplock
   Note: In all of these throwing maneuvers it is possible to use either the hips as a fulcrum or dropping to one knee and pulling opponent over the shoulder.

Lesson #26. Disarming, club or knife (Chapter 10)
1. Club
   a. Unarmed
      (1) Cross wrist block, flying mare
      (2) Outward arm parry
      (3) Forearm parry, to reverse double wristlock, and hiplock
   b. Armed with club or similar weapon
      (1) Club parries
      (2) Offensive club maneuvers
2. Knife
   a. Unarmed

(1) Downward thrust
(2) Sideward thrust, use -V"
(3) Upward thrust, use "V," double wristlock, or forearm block to double wristlock

b. Armed with dub or similar weapon
(1) Parries and offensive attack

3. Offensive maneuvers with club and knife (vulnerable and vital areas)

Lesson #27. Disarming maneuvers (Chapter 10)
1. Pistol. (Automatic or revolver)
A short lecture by the instructor calling attention to the construction peculiarities of the revolver and the automatic

a. Shoulder or cross draw holster
(1) Reverse wristlock
(2) Face strike
(3) Check hand with pistol in holster kneelift to groin or kick

b. Side holster
(1) Twisting hammerlock
(2) Double wristlock

c. Gun drawn (hands raised)
(1) Inside turn
(2) Outside turn
(3) Scissor swing (gun in face)
(4) Scissor swing (gun in stomach)
(5) Emergency (gun at side), shoulder grasp, reverse wristlock

d. Gun drawn hands down
(1) Up and down scissor
(2) Inside turn
(3) Outside turn
(4) Arm strike (gun at temple)

e. Gun in back
(1) Outside turn, lock arm, kicks
(2) Inside turn, lock arm, kicks
(3) Gun high in back, outside turn double reverse wristlock

- f. Proper use of pistol or automatic
  - (1) Keep distance
  - (2) Gun at side, hip high

Lesson #28. Disarming maneuvers, rifle. These maneuvers should be worked first without a bayonet, later a bayonet should be used in all lessons. (Chapter 10)

1. Facing opponent
   - a. Rifle twist, release and strike opponent with butt
   - b. Rifle strike and grasp, attack opponent, kicks or blows
   - c. Snap bayoneted rifle into ground
   - d. Rifle strike, twist, hiplock opponent
2. Rifle, in back
   - a. Outside turn, lock and counter kicks
   - b. Outside turn, twist rifle free, stand out of range
   - c. Inside turn, lock and counter kicks
   - d. Inside turn, twists rifle free
   - e. Rifle twist from elbow strike
   - f. Outside turn, hiplock
3. Proper use of rifle
   - a. Open-country carry
   - b. Closed-quarters carry
   - c. Taking prisoners off carry at side (hip high)

Lesson #29. Offensive methods of subduing or killing an enemy (Chapter 11)

1. Proper use of pistol or rifle
2. Most effective use of a club
   - a. Check club maneuvers
3. Most effective use of a knife
   - a. Check knife maneuvers—jabs, thrusts, and slices to vulnerable areas
4. Wire (barbed wire) or cord strangle
   - a. Barbed wire drag across face
   - b. Wire or cord loop to neck, jerk and twist ends
   - c. Two hand loop around neck, jerk and twist ends
5. Attacking from the rear
   - a. Kicks to spine, hock of knee, or tail bone (coccyx)

b. Clothing strangles, kneelift to back (snap spine)
c. Double locked strangle, kneelift to spine
d. Full nelson, trip forward, drop to deck head first
e. Grasp across face-gouging eyes and ripping nostrils
f. Leg pick up, drop to deck on head
g. Ear clap

6. Opponent on deck
    a. Shoulder and armlocks, force joints
    b. Knee drops to vital and vulnerable areas
    c. Knife, jabs, thrusts and slices to vital and vulnerable areas
    d. Hand blows to vital and vulnerable areas
    e. Heel and toe kicks to vital and vulnerable areas (especially spine)

Lesson #30. (Time allotted by officer in charge) This last lesson should serve as a general review with particular emphasis on the most effective counter-maneuvers for each position of close-in combat.

Index

U

V

W